Swampgas Theory

(by Chris Arteno)

Part 1 – Theory of Perpetuality

(The Reality of the Universe)

Part 2 – Swampgas Theories

(Some other Out of This World Theories)

Swampgas Theory

By Chris Arteno

Dedication

This book is dedicated to the provider of all possible information, and the creation of time. Without these two things, our existence would be, nonexistent!

A secondary dedication, goes to the entities of "Religion" and "Science." May they learn to exist together, for the betterment of our understanding of all that is. This coexistence must happen, in order for humanity to take the "Next Step"!

Prologue

- New beginnings, are often looked upon with ridicule, and labeled as lunacy. The truth does not require approval. It is, and will always be, the truth!
- Be open to the ridicule, and to the lunacy, that leads you to the truth. Your true path to knowledge depends on it!

Contents

Preface

Did the Universe create reality, or did reality create the Universe?

It's kind of a "chicken and the egg" sort of question. Either way you look at it, The Universe and reality have a completely symbiotic relationship. Two sides of the same coin. The Universe does not exist without reality, and reality does not exist without The Universe.

This book is written with the "Hope" of science proving it right, or proving it wrong. If it's right, we know. If it's wrong, we need to get it right.

To get it right, we need to look for the answers, with a 100% totally open mind. Closed minds lead to unasked, unanswered, and incorrectly answered questions. In this scenario we can never know "The Truth."

Introduction

There are many theories about how The Universe works. These scientific theories deal with how things work inside The Universe.

The "Theory of Perpetuality," is a look at the Universe from the outside looking in. From this "outside view," we will look at the reality of The Universe, and how reality and The Universe work together. We will also look at some of the inside the Universe stuff, as it pertains to the reality of the Universe.

Mankind has been trying to make sense of the Universe, since before we even knew there was a universe. We've always wanted to know how things work. We have travelled lightyears in our knowledge, yet we still have a very long way to go. Since the industrial revolution, our understanding of The Universe has come a long way. Currently, our knowledge is growing almost faster than we can comprehend it.

With all this knowledge, there are still a lot of things that don't make sense. They just don't add up. With so many unasked and unanswered questions, and so many unknowns, we need to take a step back, and reduce the Universe to its most basic form. Only then, can we gain a better understanding of what The Universe is, and how it works.

This book is a feeble attempt to pass on my knowledge, and my understanding of how The Universe works.

The resources used in this book, include information from both the Religious, and Scientific communities, as well as my own conjecture and opinions, but most of all, the use of Universal Knowledge! There may even be a little bullshit sprinkled in for effect.

*****all other theories are crap! *****

Part One – The Theory of Perpetuality

(The Reality of the Universe)

*** all other theories are crap! ***

Beginning "Crap"

Question

If a tree falls in the woods, and no one is there to hear it, does it make a sound?

Answer

Of course it does. It absolutely does. It 100% for sure makes a sound. Our existence has nothing to do with how Reality works! And who cares anyway? "We don't even know how The Universe works."

If you keep things simple, they might make more sense!

The Need to Know

My personal journey for knowledge, brought me to the point where I needed answers. For many of life's (big) questions, the answers available were not adequate, or there were no answers at all. When I say life's big questions, I'm talking about how The Universe works. So many questions, so few answers. Many of these questions lead to even more questions.

There is literally not enough time to learn everything I wanted/needed to know about The Universe (especially since my interest in it didn't pique until I was fifty years old.) Finding some answers would definitely bring me some peace of mind.

I asked myself, what is it that you want to know? With a little bit of time and thought, I came up with an answer. I want to know how The Universe works! Just the big stuff. Not necessarily all the grandiose science and details about everything inside The Universe (although learning some of that stuff along the way is definitely ok). I need to start by keeping it simple. I like simple! I want a simplified, big picture look at The Universe, and how it works.

Now it's time to see if "we" can come up with that simplified look. When I say "we," I am referring to myself and to The Universe itself, in the form of universal knowledge. Again, there may be a little bit of bullshit mixed in along the way.

In this chapter we are going to look at some of the things we want to know about. Also, we will talk about some things we can forget about (temporarily). Most importantly, we will talk about getting answers by tapping into "Universal Knowledge."

Let's start by forgetting about some of the things we already know (or think we know) about The Universe. We will not be talking much about Relativity, or Quantum Mechanics. Why forget about the most important sources of information about The Universe that are available? The answer to that question is: We need to start with a clean slate, and an open mind, in order to see The Universe from a broader (outside the box) perspective, with a simpler look.

We cannot get the answers we need, if we limit ourselves to Relativity and Quantum Physics. There is much more to it than that!

So, what specifically do we want to know?

- How did The Universe begin?
- How will The Universe end?

- What is Reality, and what is The Universe's roll in it?
- What is Time? (this is a big one)
- Is there a GOD? (also, kind of a big one)
- Can we come up with a simpler overall model of The Universe?

There is a lot of other "crap" I would like to know about, but this seems like a good place to start.

NOTE

***There is one other "Big Question" out there that we will never know the answer to, and that question is: "What is the purpose of it all?" ***

Now that the questions are out there, how do we find the answers? Some of the answers can be found through science and religion. Some may just be common sense, or even personal opinion. For me, most of these answers would come (in part) by tapping into "Universal Knowledge."

"Universal Knowledge!" What is that? What does it mean, and how does it work? Great questions. I can only speak on how it works for me. What is Universal Knowledge? Simply put, it is recorded knowledge that is available for anyone to (possibly) tap into. The Universe will not give you more than you can handle, or comprehend.

How does it work? For me, it started in that place somewhere between being awake, and falling asleep. In this very relaxed state, I would think of a question that I wanted an answer to. Then I would wait for the answer to come to me (through Universal Knowledge.) I would sometimes have to think about the answers multiple times to make sure I was understanding them.

Many of these answers, were things I had never heard of, or thought of before. Also, many nights I would wake with some "New" knowledge, and have to write it down on a tablet (so as not to forget it) that I kept on the nightstand. Many times, I would look at what I had written down from the night before and think, WTF is this?

How do you know these are just not your own thoughts? When it happens to you, you'll just know! But that wasn't good enough for me. I would look at this new information and think: Is this something I would have thought of on my own? and would I have understood these things on my own? The answer was, and is always no.

After months of tapping into Universal Knowledge during these "sleep zones," it started getting easier. I wouldn't have to write everything down, or re-ask the same questions more than once. I would however, continue to think about the information I was receiving, as most of the information would lead to more questions, and more answers. Months later, I would be able to "connect" to the Universe while completely awake. Maybe while relaxing after work in the reclining chair. Later, I could connect while doing other things, like listening to music, or even while working, or driving.

At some point, I would no longer have to ask questions, I would just get answers. The Universe seemed to already know the questions I had, and just provided the answers (kind of scary.)

In summary, we need some answers, so let's see if we can get some!

We might as well start with the biggest question first.

What is The Universe? That really is a big question. Let's see what the simplified answer is. The answer will undoubtably result in more questions.

Question: What is The Universe?

Answer : The Universe is Reality.

Question: What is Reality?

Answer : Reality is Information attached to Forward Moving (stable) Time.

Question: What is Information?

Answer : Information is any Energy/Matter Relationship.

To break it down again:

Universe = Reality or U=R

Reality = Information attached to time or

$R=(em)^r +T\rightarrow$

This is the simplest definition of what The Universe is!

We will spend more **time** on this later.

NOTE

Any answers used to explain the Universe, cannot result in any form of a "Paradox."

***all other theories are crap! ***

<u>*Between Chapter "Crap"*</u>

"Paradox"

"Any question who's answer results in a paradox, cannot be accepted as proof to that question" ... a paradox is defined as, a contradictory statement that may result in an unlikely truth. Since a paradox "may" result in truth, it stands to reason, that it may also result in an untruth! With the possibility of an untruth in any answer, the possibility of truth is now tainted or diluted.

"Paradox"

What a great name for a rock band! It's such a great name, that it has been used at least three times in the history of "Rock N Roll." The first occurrence was in 1986 by a "Thrash Metal Band" from Germany. Also, from the 1980's, a Canadian band brandished the name of Paradox. Finally, in 2019 an alternate rock band from Australia used the name "Paradox" (like I said, what a great name for a rock band.)

In The Beginning

Chapter 2

Question : Did The Universe always exist?

Answer : No, it did not!

Before we look at the beginning, let's look at the Universe before the beginning. Was there anything before the beginning? How could anything exist before anything existed? As it turns out, a couple of things did exist before anything existed.

The first of which, and the most important, was this little thing we call "Time."

Yes, time existed before the Universe existed.

*** This is where the "Theory of Perpetuality" differs from most, or all other theories." ***

As a matter of fact, without time, The Universe could not exist at all.

***Note: Time is so important, that it will have its own chapter later in the book. It will most likely show up in every other chapter in the book. ***

The second thing that existed before the Universe existed, but not before time existed, was a whole lot of nothing. This whole lot of nothing, was the void that would someday become "Spacetime." In the beginning this whole lot of nothing (due to its expandability) would definitely become something!

Ok, back to the beginning, and the "Big Question." How did The Universe begin? It may not be the "Mother of all Questions," but it does come in as a close second.

Most of us categorize the answer to this question, as having two distinct possibilities. There are however, three possible options that we will talk about in this chapter.

Saying that "nothing but time," or "a whole lot of nothing and time," were the only things that existed before the Universe did, is not something you can prove. This is why it's called a theory. Now, let's check out the three options.

- Option One: The oldest of the options, and probably the most accepted of the three, is called "Creation." This means that the Universe, and everything in it, was created by a *Supreme Being*, a *Higher Power* or by *GOD!* In this option, science, and chance, have nothing to do with the creation of The Universe, or anything in it. Let's explore the "Creation" option a little further.

Why do so many believe this option? The question of creation has been around for almost as long as humanity has been. In the early days of humanity, religion was the only source available to provide answers to life's important questions. For many, religion is still the best place to find these answers.

What does religion say about creation?

Genesis:1 – In the beginning, God created the heavens and the earth (notice the "s" on the end of the word heaven).

Genesis 2 – And the Earth was without form and void.

When The Bible says God created the Heavens and the Earth, it means He created the realm of "Heaven" where he resides, as well as any other realms (including the realm of time) that exists outside of our universe. The word Earth, is synonymous with the word universe. So, God created our universe, and all the possibilities that could exist in our universe. The Earth (universe) was without form and void. This means that before the beginning, the only things that existed were, God's Heaven, Time, and the Universal void.

There were no possibilities, or possible information, in the void at this time. In the beginning, "Information" or "Energy/Matter relationships," were introduced into the void. Since time already existed in the void, the introduced information was attached to time, and the void was transformed into what we call "spacetime" (More on spacetime later.) What we call the Universe, Spacetime, and Reality, are now all the same thing.

To summarize Option One; God Created everything inside and outside of our universe. This includes infinite time. This also includes all that is known, and all that is not known. As far as our universe is concerned:

- God created a large (expandable) void, where only time existed.
- God then created all the possibilities that could exist in the Universe (i=em) ^{r}. Remember the definition of information is any Energy/Matter relationship.
- God allowed all possible information to enter the void, where it would attach to forward moving, stable time. This would transform the void into Spacetime. All possible information will remain attached to time, in the form of past or present time, for as long as The Universe exists.

Note

In The Bible, God refers to himself as "The Great I Am."

Is it possible that the original Hebrew translation of this was incorrect? Could God have been saying: I am Great, I am all possibilities, or I am the creator of information $\{i=(em)^r\}$ or, with God, all things are possible?

If Option One is true, that means there is/are Omnipotent, Supreme Being(s), that exist in another, or in many other realms. These beings can transcend space, and time. These "Supreme Beings" do not exist in our universe, or any other universe. They are from a realm, that allows them to travel through space and time, with ease, and with speed. They may even be able to exist everywhere at the same time, or everywhere all of the time. They may be from the realm we call "Heaven." (We'll talk about other realms later in the book.)

Those who believe in the theory of "Creation" (and it is a theory, as it cannot be proven,) dismiss Science as having any part in the creation of the Universe. This is clearly a mistake. One that leads to having closed minds, and to having unasked, unanswered, and incorrectly answered questions. Science must play a part in any theory of creation. A failure to do so will lead to a non-working model of The Universe, and how it works.

"Creation is a completely faith-based theory!"

"But wait, there's more." Let's move on to Option Two of how the Universe began.

- Option Two: While being a much newer theory, it is also a very popular theory. Option Two is known as "The Big Bang Theory." What is The Big Bang Theory? (Much more than a TV show). Simply put, The Big Bang Theory, says that The Universe, and everything that it would become, was condensed into an unimaginably small and dense point called, a singularity. This singularity exploded (not in a fiery violent way), pushing everything into existence at the fastest rate possible.

The Big Bang Theory, is a completely scienced base theory. In this option, the Universe/Reality/Spacetime, is created from a singular and unimaginably dense point that existed somewhere inside or outside of the void, that would become Spacetime.

Yes, The Universe began with an explosion of Energy and Matter. (QGP) would eventually give way to Hydrogen, and the rest as they say, "Is History!" The Big Bang Theory, just as its Creationary counterpart, cannot be proven, and is also just a theory.

Also, just as in the Creation model, spacetime is created when information is introduced into the void, and is attached to forward moving, stable time.

Believers in The Big Bang Theory, dismiss religion as having any part in the creation of The Universe. Clearly a mistake, as any version of the origin of The Universe that does not include religion, will be invalid!

Those who believe in the Creation model, may believe the Big Bang model is ridiculous. Everything just appearing out of nowhere, then exploding into what we now call the Universe (sounds made up.)

To look at everything that The Universe is, from galaxies to stars, and planetary systems, all the way down to the smallest particles in existence, and to think, it all just happened, without some sort of a "Grand Design," doesn't seem possible.

The believers in The Big Bang Theory, believe the Creation model to be equally ridiculous. Everything that is our universe, being created by another living being, or beings? How can that be possible?

The creation model, also leads to many more questions.

- Who are these higher life forms?
- Where do they come from?
- Did someone create them?
- Why do they not reveal themselves to us?
- Why did they create everything?

The Big Bang Theory, also leads to more unanswered questions.

- Where did the singularity come from?
- What made it explode?
- Could there have been more than one singularity?

Since neither option can be proven, and neither can lead us to a clear working model of the Universe, what else is there? Is there an answer we can all live with? What explains how it all started? This leads us to Option Three. Just as Option One, and Option Two cannot be proven by science, neither can Option Three. Option Three is not completely Faith Based, nor is it completely Science Based.

Option Three, is a "Best (educated) Guess," guided by logic and common sense, with a little bit of proof to back it up. Option Three, might be the option that is as close to the truth as we will ever get. Let's take a look at Option Three.

- Option Three: Since neither Faith or Proof, can supply a complete working model of how The Universe began, we must use a "Blended Model" of both option One, and Option Two. This "Blended Model," will require some "Faith," as well as some "Proof." Option Three will be known as "GO."

Just as in Creation and The Big Bang, with "GO," information is introduced into the void, and is attached to forward moving (stable) time. In "GO," the release of information is the result of a Supreme Being(s), "Gifting all possible information into the Universe. Everything that could ever happen in the Universe is released, with a single word from *GOD!* "GO!"

This "Gifted Information," enters the Universe, as the result of a "GOD created, Big Bang," known as "GO." From that point on, it's all science! The Universe is allowed to unfold naturally, without any outside influence. Does this mean that the outcome of the Universe is predestined? The answer to that depends on whether or not you believe in "Free Will." (We will look at "Free Will" a little later in the book.)

To recap, Option Three says that GOD created the Universe, and with a single word (GO), set it into motion, then took a step back, to let The Universe, become The Universe! Everything did not appear out of nowhere. There was/is, a "Grand Design." However, GOD does not control the destiny of the Universe. The difference between Option Two and Option Three is: What caused the beginning. In Option Two, we don't know. In Option Three, we do!

The Universe had some help getting started. Now it's up to science to figure out how it all works!

Universal Knowledge, leads me to believe that "GO," is the best option available, and the option that is closest to the truth. Option Three requires 100%, absolute faith, and strives for 100%, absolute proof.

***all other theories are crap! ***

Between Chapter "Crap"

"Narcissism"

(I hate that word, and it's so hard to spell!)

A man, who is a self-proclaimed narcissist, once said to me, "All roads lead to narcissism." What was he saying? I think he was saying that everyone, to some degree, is a narcissist. He offered the argument of self-preservation, as proof to his theory.

My "Knee-Jerk" reaction was to say, "you are wrong," and "you're an idiot." However, being a man with an open mind, I spent a fair amount of time thinking about this. The first thing I did was to look up the definition (and the correct spelling) of narcissism. Here is the definition that I found. "An excessive interest and admiration in one's self. A selfishness involving a sense of entitlement, and a lack of empathy. An extreme need for admiration."

***This is actually characterized as a disorder. ***

After some more thought, I realized that there are definitely those who meet these criteria. There are also those among us who are completely selfless! Most of us, fit somewhere in the middle. Due to the fact that there are some completely selfless people in the world, his statement that "All roads lead to narcissism," is not correct.

As a species, I believe we are narcissistic! We have a "Me First," mentality. This is a mentality, that must change!

This is when I first realized that everything has a polar opposite (selflessness and narcissism).

"You stole my possibilities, to feed your narcissistic hunger, You selfish, selfish bastard!"

It's About Time!

Chapter 3

Time! Since the beginning of time, man has been obsessed with it, fascinated, and intrigued by it. As of yet, we have never fully understood it!

We briefly talked about time in the previous chapter. How time existed, along with the void, before The Universe itself came into existence. How can something exist, before anything existed? To answer that, we need to answer another question, a really big question. What is time?   Ok, so what is time?

- Time is a measurement.
- Time is constant.
- Time is infinite.
- Time can be burdened, or unburdened.
- Time is a physical property of reality.
- Time is real.

Before we take a more in depth look at what time is, let's look at what time is to us. None of these things, are what time really is, but how time is perceived in our daily lives. For us, time can be many things.

- Time is a memory.
- Time can help us heal.
- Time is a deadline.

- Time brings order to our days.
- Time can be regret.
- Time can be a dream.
- Time is a race towards death.

Yes, time can be many things to us. These things are all relative. However, time is not relative. Time is not perception, and time is not a man-made concept! Most of all, time cannot be altered! I will attempt to address each one of these items (hopefully without going on a rant.)

First off, time is not relative. Relativity, in regards to time, will leave any model of the Universe in a complete "Paradox" (Sorry AE.) Secondly, time is not perception! Although, the human mind definitely has a perception of time (Sorry BB. Don't worry, he knows who he is.) Next up, time is not a man-made concept. Time is very real. As a matter of fact, time is so real, reality would not exist without it. Lastly, no explanation needed on this one, time **cannot** be altered!

Enough of that, let's get back to what time is.

→ *Time is a measurement.* Time is a tool, that we use to guide us through each day, and through life's many events. Throughout our lives, from beginning to end, we depend on time more than any other thing that is in existence, without ever giving it a thought. Yes, time is a measurement, but without it, we could not exist.

→ *Time is constant.* Time is constant, and as a measurement, it should be. Now this is where I'm going to get myself into trouble. Since time is constant, it cannot be altered in any way, shape, or form. Sorry "Relativity," you've got it wrong (kind of.) Time does not change! Ever! It does not speed up, and it does not slow down. Time cannot be changed by gravity, velocity, or the curvature of spacetime. What does change, is the way that information is attached to time. This is due to the "Temporal Velocity," and the "Temporal Adhesion" of information changing, in regards to time.

→ *Time is infinite.* This can be a tough concept for us to understand. Since time is a measurement, and time is constant, it seems logical that time would also be infinite. But what does infinity mean? For time, it means this: There is no beginning, and there is no end. Also, there is this little thing we call the present. The present can be measured as one Planck time. BTW, in this book, we will be replacing the term Planck Time with the acronym SMAT (Smallest Measurable Amount of Time.) No disrespect intended to Max Planck. I just seem to do better with the term SMAT. Not only does time have no beginning, or no end, it is also all encompassing. Time is everywhere, or omnipresent. Time knows no boundaries. Time exists inside our universe, and it exists outside of our universe. It exists in all other universes and everywhere in between. Time existed before our universe ever did, and it will exist long after our universe is gone. Time is truly infinite. Not only is time infinite, time is the only thing that is infinite. Wait, the Universe is not infinite? Nope, it has a beginning and it will have an end! What about alternate universes? Those don't actually exist, but if they did, they would be finite. Alternate realities are also finite. Imagining a forever future, is not difficult for us to understand. Time just keeps marching on. It's the forever past that seems to give us some problems. How can something have always existed? How can something have no beginning? Those are some tough questions!

→ *Time has properties.* When I Googled time, to see what they thought of it, I kept getting results for Thyme. I just kept thinking, "Hey Google," this isn't a cook book! So, what are the properties of time that I am talking about? There are two major properties of time. Two things that time can be. The first thing is, time can be burdened. The second is, time can be unburdened. What is burdened time? Burdened time, is when information is attached to time. This is how time works while it's inside our universe. Burdened time is how reality is formed.

Information has no part, in the process of becoming attached to time. Time is in full control of that process. (Not sure if that's important, it is however the way that it is.)   *** *** ***

Unburdened time is simple. Unburdened time, is when time exists, without any information being attached to it. This is how time exists, outside of the Universe. These are the two major properties of time.

→ _Time is Real._ I can't say much more about this, other than to say Time is Real!

*** _all other theories are crap!_ ***

Between Chapter "Crap"

"Time" (the playlist)

• Turn, Turn, Turn	(The Byrds)
• Time	(Pink Floyd)
• Time is on my side	(The Rolling Stones)
• Yesterday	(The Beatles)
• Does anybody really know what time it is?	(Chicago)
• Time of the Season	(The Zombies)
• Cats in the Cradle	(Harry Chapin)
• Lady	(The Little River Band)
• Back in Time	(Huey Lewis)
• Time Marches On	(Tracy Lawrence)
• Time in a Bottle	(Jim Croce)
• Realin' in the Years	(Steely Dan)
• Time Passages	(Al Stewart)
• Good Times Roll	(The Cars)
• Time	(Alan Parsons Project)
• Too Much Time on My Hands	(Styx)
• Time After Time	(Cindy Lauper)
• Time Has Come Today	(Chambers Brothers)
• No Time	(The Guess Who)
• Fly Like an Eagle	(Steve Miller Band)
• Hair of the Dog	(Nazareth)
• I am I Said	(Neil Diamond)
• The Rockin' Years	(Dolly Parton)
• Grandpa	(The Judds)
• Brother Jukebox	(Mark Chesnutt)
• Cool Change	(The Little River Band)
• Bubba Hyde	(Diamond Rio)

Who Needs Information?

Chapter 4

In this Chapter, we will take a closer look at information, and its role in The Universe. Let's do a quick review of what information is. The definition of information is really very simple. It's any energy/matter relationship. It doesn't matter how big or how small these relationships are. It doesn't matter what the ratio of energy to matter is. All energy/matter relationships, are considered to be information.

We will now divide information into categories, and sub-categories. Also, we will explore what happens when information interacts with time.

OK, so who does need information? or to be more accurate, what needs information?

- Spacetime needs information.
- Reality needs information.
- The Universe needs information.

The reason Spacetime, Reality, and The Universe all need information is... They are all the same thing. The Universe is Reality, being played out in Spacetime (this is important.)

Let's jump back, to just before the beginning of the original, or very first universe. When the first universe was created, (Oh! that must mean that there is more than one universe!) all information was gifted to The Universe (which, at that time, was only a void) by its "Creator."

At the time of creation/the big bang/go, all information was released into the void (where time already existed) where it was attached to time.

***Note – this book, is where the term: *Temporal Adhesion* is first introduced. ***

This attachment of information to time, inside the void, is how spacetime was formed! This is where the categorization of information begins. When information was first introduced into the void, it was in the form of "All Possible Information," or "Possibilities" (this is the main category of information.) At that point, nothing existed, except possibilities.

At "GO," All Possibilities split into two sub-categories. These sub-categories, are "Actuality" and "Impossibilities." Actuality must begin, before any Impossibilities can be created. As actuality begins, (that just means that things in our reality start to happen) the things that are happening cause other things, to NOT be able to happen. An example of this would be: If you get out of bed at 7:00 a.m. on any particular day, getting up at any other time would no longer be possible. These things which can no longer happen, become impossibilities.

The sub-category, Actuality, will have a sub-category of its own. This sub-category is known as Actualities Past. (Some of you, may have already guessed this.) We will shorten this to "The Past." The sub-category called Impossibilities is a dead end. This is the end of the line for any information that ends up here. Once the All-Possibilities category splits, it will become "Remaining Possibilities." For future reference, these will both be known simply as, "Possibilities."

There is one more split to talk about. That is when the sub-category, "The Past," splits into two sub-categories of its own. The Past will split into the sub-categories called the "Grey Past," and the "Dark Past."

These are the Categories that information breaks down into. We will now list them in "bullet form," for a better visualization.

- Possibilities (All Possibilities/Remaining Possibilities)
  - Actuality
    - The Past
      - The Grey Past
      - The Dark Past
  - Impossibilities

All of these categories, and sub-categories, are a part of our universe's reality.

Now let's set it all in motion. Possibilities enter the Universe, and become attached to time. These

Possibilities then become either Actualities or Impossibilities. As Actualities happen, they cause Impossibilities to happen as well.

Impossibilities remain attached to time, at the moment they became an Impossibility. Impossibilities can no longer become an Actuality.

Actualities (things that are happening in our reality) are what we call The Present. The Present happens one SMAT at a time, then very quickly becomes The Past (everything you do, instantly becomes the past.) There is not much more to say about The Present. It's here, it's gone!

Let's move on to The Past. This one's pretty simple too (or is it?) The Past should be an easy concept to understand. Not so fast, "The Past" has two sub-categories of its own. The First of which is The Grey Past. The Grey Past is any information (that is attached to time in the moment that it became The Past) that can affect the probability of any Possibility becoming Actuality. Once the Grey Past has no probability of affecting The Future, it will become The Dark Past! There will be no memory of, or possibility of The Dark Past ever being rediscovered. The Dark Past now exists only as a historical recording in The Universe.

Any past information event that goes straight to the Dark Past (bypassing The Grey Past entirely) is known as a "Null Event." Null events are very rare. ***

Now, we can talk about The Future. Wait a minute, The Future isn't even on our list of categories and sub-categories of information. There is a good reason for this. "There is no such thing as "The Future!" That's right, no such thing as the future. How can there be a present if there is no future? What happens in The Present, is a result of information from The Grey Past, and Information from the All-Remaining Possibilities category, coming together to form the most probable future. This most probable future is what becomes The Present!

Ok, there is no future, but there is a very large number of possible futures. There is also another factor, that can affect the process of information becoming The Present. This "Factor" is called "free will." We will talk more about "free will" later in the book. Also, later in the book, we will talk about the possibility of "Impossibilities" NOT being at the end of their road.

NOTE All possibilities, will eventually become either an actuality (and then an actualities past) or an impossibility. At that point, there will no longer be any remaining possibilities, and The Universe will end. Because the number of possibilities is a very large but finite number, reality, and The Universe itself must also be finite. There was a beginning to The Universe, and there will be an end to The Universe. ***

    To recap:

- Information is any Energy/Matter relationship.
- Reality is information attached to forward moving, stable time.
- Information is broken down into categories, and sub-categories.
- Reality is played out in Spacetime.

***all other theories are crap! ***

Between Chapter "Crap"

"Swampgas Rules"

- Don't worry about things you can't control.
- Control as much as you can.
- Set yourself up "not to fail."
- Have a plan, but don't be afraid to change it.
- Don't care about what other people think of you.
- Always be yourself, and be happy with yourself.
- Sometimes, it's ok to just let things happen.
- Pick and choose your battles, know when to walk away.
- If you make a bad decision, it's ok to forgive yourself.
- Remember, the grass is never greener!
- It can't hurt to ask, so ask!
- A little bit of "crazy," is not a bad thing.
- The truth is full of lies.

Spacetime "The Final Frontier"

Chapter 5

Spacetime, what is it, what does it do, and how does it work? In this chapter, we will try to answer these questions. We already know that spacetime began as a void. Time was the only thing that existed inside of the void. When (gifted) information was introduce into the void, it was immediately attached to time. This is how spacetime was formed.

So, spacetime is the place where information resides. Information does not just reside there, it (with the help of time) becomes reality there. Spacetime is the place where reality is played out. Spacetime also has another very important function.

Spacetime's second, but equally important function, is to create a historical record of information's journey through reality. This historical recording, includes all phases of information, from Possibilities to Actuality, to Actualities Past, and lastly, Impossibilities. Spacetime, somehow records the entire and complete history of The Universe (Wow!)

Spacetime becomes the place where reality is played out, and also where the game of reality is recorded. How does this work, and why have we never seen evidence of this recording of information? The historical recording of The Universe, happens in a place in The Universe that we do not have, or have not found access to. This place is either sub-spacetime, or inverted spacetime.

Let's recap spacetime in bullet points, and see if it all makes sense.

- Space exists as a void.
- Time exists inside and outside of the void, but has no relationship with the void.
- Information is introduced into the void.
- Information is attached to time in the void.
- Time is converted from unburdened time, to burdened time.
- Burdened time (which includes its attached information) combines with space (previously the void) to become spacetime.
- Spacetime becomes the place where reality is played out.
- Spacetime creates a historical recording of how reality is played out.
- As possibilities run out, reality ends.
- Spacetime collapses into a historical singularity (which includes the original beginning possibilities, from the beginning of The Universe, to be passed on to the next universe.)

- Spacetime, Reality and any remnants of The Universe will come to an end. It will be as if they never existed.

*** Spacetime Notes***

- Time, Space, and Information, have a truly "Triunionally, Semi-Obligative, Symbiotic relationship. Time and space cannot become spacetime without each other. Without spacetime, information has nowhere to exist. Without information, spacetime has no reason to exist. Time can, and does exist, with or without information, however in this unburdened state, it cannot be realized. Without reality, time is a very lonely entity.
- Reasons why spacetime can appear to bend.
    - The shifting of information from category to category, may appear as a bend in spacetime.
    - The historical recording of information, may also appear as a bend in spacetime.
    - Any change in information's Temporal Adhesion, or Temporal Velocity, could possibly appear as a spacetime fluctuation. Temporal Adhesion, and Temporal Velocity, can be affected by gravity, or physical velocity.

*** Universal Thought***

   Since energy/matter cannot be created or destroyed, and information is made of energy/matter relationships, then information, also can not be created, nor can it be destroyed. This explains why All Possibilities must have been "Gifted." This also explains why reality, eventually becomes a "Historical Spacetime Singularity," that will always exist at the point in time, where reality became the historical singularity. In this scenario, The Universe never really ends, it just shifts, from an ongoing actuality to a permanently recorded past!

*** all other theories are crap! ***

Between Chapter "Crap"

"Make Good Choices"

Before creating your own personal reality, always ask, "Why am I going to do this?" Here are some good reasons.

- It makes you happy, without harming others.
- It benefits you, without taking advantage of others.
- It makes others happy, with no regard to your own happiness.
- It benefits others, regardless of its effect on you.
- It's the "right thing" to do!

Here are some neutral, or not so good reasons to do something.

- Just because (not a good reason.)
- Wanted to see what would happen (neutral reason.)
- Intrusive thought (a very, very bad reason to do anything.)
- I had no other choice (be careful with this one.)
- Going with the flow (usually not a good idea.)

Alternate Reality
Oh Crap! Rules Apply

(Chapter 6)

Reality, what a scary word, and an even scarier place. Simplifying reality, may be the best way of understanding it. Here is a review of the phases of reality, which is where the categories of information reside.

- Totality – All possible information.
- Eventuality – All remaining possible information.
- Actuality – All present information (1 SMAT.)
- Relevantuality – Grey Past Information.
- Finality – Dark Past Information.
- Terminality – Impossibilities or "Null" reality.

There is a lot going on with information forming reality. There is one category of reality that we haven't talked about yet. Alternate Realities, this is the "stuff" that Science Fiction is made of. But, oh crap! there are rules to alternate realities in The Universe. In this chapter, we will look at the types and levels of alternate realities in The Universe, and the "rules" that apply to them.

Before we start, we need to get some things straight. Alternate universes, and alternate realities, are not the same thing. Inside our universe, there are no alternate universes, there is no such thing! There may be other similar universes to ours, with many similarities to our own. These could possibly exist, but they would exist outside of our universe, and there are no ties to our universe.

Also, we do not create new universes, or alternate realities, every time we make a decision. Creating a new reality, or something as enormous, and complex as a universe, out of our own free will decisions, is complete and total crap! A narcissistic view, from a narcissistic race! Having said all that, there are possibilities of alternate realities in our universe (with rules.)

Let's move on with the types of alternate realities. The first type is called an ITS (If Then Scenarios.) An ITS is a preview of a possible (future) Alternate Reality. If it is necessary for an ITS to become an alternate reality, it would become a TAR (Temporary Alternate Reality.) Yes, any alternate reality is only temporary, and would rejoin the original reality's timeline at some point. TARs do not affect most of The Universe, and will disappear from any future reality.

Here are the three levels of TARs.

1. Localized
2. Generalized
3. Critical

A Localized TAR, affects a very small portion of The Universe, possibly just a handful of people. They are short lived, and have no effect on the rest of The Universe. These alternate realities are somewhat common. "Ever wake up in the morning and feel like, somethings changed, without knowing what?" (Localized TAR.)

A Generalize TAR, is similar to a localized TAR, but has a larger effect on The Universe. The portions of The Universe that are affected, might be on a planetary, or galactic scale, or even affect galaxy clusters. These TARs will also reconnect to the original timeline, without any effect on the outcome of The Universe. The Generalized TARs, are not as common as their localized cousins. They are somewhat rare.

Critical Alternate Realities (this is the Oh Crap! one) would change the outcome of The Universe. This alternate reality would forever change how The Universe unfolds, and ultimately when and how the Universe would end. If a Critical Alternate Reality were to happen, we would have no knowledge of it. Our future would change from the original timeline's future. Our past will not have changed, but the memory of our past, will be a very different one. The chances of a Critical Alternate Reality happening, are "Ultra Rare," and would most likely never happen.

Now that we know the types, and levels of alternate realities, let's look at the causes of them, and the rules that apply to them.

First, let's look at the causes of alternate realities happening.

1. ITS (If Then Scenarios)

An ITS occurs at the point where Possibilities become Actuality. The Universe will "run the math," to see which ITS has the best probability, of becoming the best actuality, as it pertains to the outcome of The Universe. The information event, with the best probability, will become Actuality. The rest will become Impossibilities (they are just not good enough.)

2. TARs (Temporary Alternate Realities)

Every once in a while, (when running the math) The Universe will come across multiple information events with the same probability of becoming the best possible Actuality. The Universe will allow for these events to all exist as actuality, at the same time. The actuality that we live in, we call Reality. The others we call TARs. At this point, our actuality is also a TAR, but we don't see it that way. TARs always begin as localized events.

3. Generalized TARs.

The reasons that Generalized TARs exist, are very different than those that cause Localized TARs and ITS.'

Here are some of those reasons.

- Interfering with or creating Black Holes.
- Terraforming Planets.
- Altering the course of asteroids and comets.
- Altering the life of a star.
- Interspecies or alien relations (physical and conceptual.)

Creating Generalized TARs can be dangerous (not to The Universe, but to its unfolding.) Interfering with The Universes natural unfolding, should not be taken lightly. There is a fine line, between learning about how The Universe works, and affecting how The Universe unfolds. There are consequences for creating Generalize TARs. We will talk more about these in a moment.

4. Critical Alternate Realities

Critical TARs, are when Generalized TARs cannot rejoin the original timeline on their own. At that point, The Universe will attempt to forcefully rejoin the two timelines. This is what is known as a "URC," or a Universal Reality Correction. If a URC is unsuccessful, a Critical Alternate Reality would occur. This would have a profoundly negative effect, on the unfolding and outcome of The Universe. The Universe would end prematurely!

*** Warning***

This is where crossing the line, has consequences!

If humanity, or any other beings, would create a Generalized TAR, with a probability of over 50% of becoming a Critical TAR, a URC may be initiated by The Universe. The elimination of humanity, may be the correction used to prevent the GTAR from going Critical!

Just as "Mother Nature "will not put up with our "Shit," neither will The Universe. The Universe will protect itself, at all costs.

Enough of all that nonsense. Before we get to the rules of reality, let's first talk about why there are rules to reality. This cannot be proven, but I believe this to be one of the biggest "Truths," out there!

The Universe is Alive! and it has "Free Will" (although its free will is somewhat limited, and could be described as "Autonomic." Some of the terms used to describe life, can also be used to describe The Universe. Here is a list of those terms.

- Growth
- Movement
- Excretion
- Metabolization
- Reproduction
- Ability to respond to its environment.
- Free Will

The Universe creates, and gives all things needed to form life. He who creates life, is life! The Universe is the highest form of life in our reality. "GOD" is a higher form of life than The Universe, as GOD is the "Creator" of The Universe. Also, GOD is not a part of our reality, He is the creator of it! GOD has His own realm, yet He can exist in any realm.

Now we can get to the rules of reality.

- Reality is under the direct, limited control of The Universe, and can be altered by The Universes free will.
- ITS' and TARs cannot be compounded. An ITS cannot have an ITS of its own. Also, a TAR cannot have a "spinoff" TAR or TARs linked to it. Compounded alternate realities, can lead to frayed realities, which would be unstable, and intolerable to The Universe. Unstable realities would lead to the eventual derailment of reality, and are not allowed in The Universe.
- TARs will always return to their original timeline (with the exception of an uncorrected critical TAR.)
- Timelines must be sustainable, and have highly desirable outcomes. Sustainability must be well over 99%.
- Events in our reality can also happen in a TAR at the same time. This is what "Deja Vue" is!
- When a TAR becomes unsustainable or unnecessary, it rejoins the original timeline.

*** Reality Thought***

Sustainability, of information in a timeline, is proof that we are all special, and should place a great value on our existence!

In Conclusion: Realities and Alternate Realities are some of the hardest things to comprehend. We **must** get this right, if we have any "Hope" of learning the truth of how The Universe works.

****all other theories are crap! ***

Between Chapter "Crap"

The Localized TAR

"Practice Squad Pre-Season"

(With Extreme Rules)

Concept:

- A three-week practice squad preseason.
- Each team's practice squad, would play one game against each of its remaining divisional teams practice squads.
- The purpose of these games would be:
  - o For Players to gain actual game experience.
  - o Elevate the level of play to that of the professional roster players.
  - o Showcase players talents, for possible promotion.
  - o Ensure call-ups to be virtually seamless.

Extreme Rules:

- Timeclock does not stop for incomplete passes, or out of bounds.
- Each team gets one timeout per quarter.
- Each team gets one injury timeout per half.
- If a team wants to stop the clock beyond these timeouts, they must give up a down.
- Kickoffs are from the 20-yard line.
- All kickoffs must be returned.

- Kicking team cannot cross the line of scrimmage until the ball is touched by the receiving team (with the exception of the kicker.)
- If the ball is kicked through the end zone, the ball will be placed at the 30-yard line.
- If a kickoff goes out of bounds, the ball will be placed at the 45-yard line.
- Punts may not be returned. They may be fair caught, or possession can be taken where the ball is downed, or where it goes out of bounds.
- If a punt enters the end zone, it will be placed at the 30-yard line.
- On fourth down, the offence may only punt or try a field goal. You cannot "go for it" on fourth down.
- Any penalty by the offence, will result in a loss of down, and a loss of five yards.
- Any penalty by the defense, will result in a five-yard penalty, and an automatic first down.
- All personal fouls result in a fifteen-yard penalty, and the player is disqualified from play for the remainder of the quarter, and the entire next quarter.
- The play clock is 30 seconds.
- Substitutions are only allowed in the first 15 seconds of the play clock.
- The ball may be snapped anywhere from 15 seconds or below, remaining on the play clock.
- Overtime is decided by 50-yard field goal attempts.

- No Challenges.
- No replays.
- Back-up refs call the games.
- All other rules remain the same.

It's Energy that Matters

(Chapter 7)

We know that energy/matter relationships, are what information is made of. When this information is attached to time, spacetime is formed, and reality is created. This reality, is then played out in spacetime. This is essentially what The Universe is.

Did you know that energy and matter play important roles in keeping order to our reality? In this chapter, we will discuss the types of energy and matter, and how they correspond to the types of information, and the phases of reality that they are responsible for controlling.

Where does energy and matter come from? Originally, they are "Gifted" to The Universe, by its "Creator." Can energy and matter be destroyed? According to the "laws of conservation," energy cannot be created nor can it be destroyed. The same is true for matter. When we say that energy and matter are two sides of the same coin, we are saying that they have a "Transformative Symbiotic Relationship." Energy and matter always exist together. Even what appears to be pure energy would have some amount of matter or mass to it. All information, "$I=(EM)^r$" has an energy/matter ratio to it. Not everyone will agree with this. It is however, the way that it is. Without each other, neither can exist.

Like types of energy and matter team together to perform certain roles in The Universe. The tasks performed, have to do with keeping categories of information, and the reality phases they exist in, from interacting, or coming in contact with other types of information, and their phase of reality. By matching the types of energy and matter, with the categories of information, and their reality phase, we can learn more about the "what's and whys" about each type of energy and matter.

We know quite a bit about normal energy and normal matter, because it's what we live in. We know a little less about Negative Energy and Antimatter. We know very little about Dark Energy and Dark Matter. We're not sure if Grey Energy and Grey Matter exist at all. And, there is something called White Matter, that is almost never talked about.

Let's put it all together. Let's match each Energy/Matter type, with each category of information, along with its phase of reality, and see what happens. Let's see if any of this makes sense.

First, we'll start with Possibilities. Possibilities do not exist in any phase of reality. Possibilities energy/matter type, is probable energy/matter, or white energy/matter. Probabilities energy/matter type, can become any other energy/matter type, depending on what type of information it becomes, and what phase of reality it converts to.

Next, we will move on to Actuality. The information type, called Actuality, exists in the reality phase called The Present. Actuality is made of, and controlled by the energy/matter type called Normal Energy and Normal Matter.

Next up, is the Actualities Relevant Past. It's phase of reality is known as the Grey Past. The Grey Past is made of, and controlled by, the energy matter group of Negative Grey Energy/Grey Antimatter. The Grey group has a ratio of Normal Energy/Matter to Negative Energy/Antimatter. The ratio is determined by the probability of the Grey Past, influencing the Present or any probable future. Since normal energy and matter don't play well with negative energy, and antimatter, a shades of grey buffer zone exists between them, in order to keep the fabric of Spacetime safe. This Grey Buffer Zone exists in Inverted Subspace.

Actualities Permanent past is next. Its reality phase is called the Dark Past. The Dark Past's energy group is Negative Energy/Antimatter.

Finally, we have the information category called Impossibilities. Its reality phase is called a "Null Reality."

Impossibilities take up the most amount of Spacetime in The Universe (something like 80-90%.) The energy/matter group for the Null Reality is Dark Energy/Dark Matter (it all sounds so ominous.)

There is a part of The Null Reality phase, that doesn't become an impossibility, until it finishes its time in an alternate reality, or its time as a TAR. This information while in a TAR situation, acts just as an actuality would. Once this information is finished as a TAR, it becomes an impossibility, but also can be classifies as a TARs grey past, and a TARs dark past. These TAR situations are governed by Dark Grey Negative Energy, and Dark Grey Antimatter.

If 80-90% of Spacetime is Dark energy, and Dark Matter, what makes up the rest of The Universe?

- The Present and The Past.
- A copy of the original All Possible Information.
- Remaining Possibilities, take up almost no space in The Universe, as they are prehistoric.
- A historical recorded copy of all information, and its journey through reality.
- There is a substance used as a Temporal Adhesive. This substance is unknown at this time. We will talk more about Temporal Adhesion and Temporal Velocity in the chapter on Time Travel.
- Time itself, takes up a very small percentage of Spacetime, as its energy and mass are temporal, and

do not follow the normal rules of physics, and cannot be measured by any means known to us.

Note

As The Universe ages, the energy/matter ratio will change from being heavy on the energy side, (closer to the beginning of The Universe) to a heavy on the matter side (the closer The Universe gets to its end.)

*** Again, when all possibilities become Actualities Past, or Impossibilities, The Universe will end. ***

***all other theories are crap! ***

Between Chapter "Crap"

"Understanding Time"

- Time and Infinity. Whatever point you are at in time, will be the half way point between the infinite past, and the infinite future. In math, infinity represents a true paradox.
- Without time, there would be no reality.
- What Religion says about Time. "There is an appointed time for everything, and a time for every event under heaven (Ecclesiastes 3:1). This is GOD telling us that reality, is information attached to time.
- What Science says about time.
    - Not absolute.
    - Fluctuating.
    - Not fundamentally real.
    - Meaningless.
    - Time is as real as space.
    - Quantum gravity eliminates time.

  Science is all over the place on this one.

- To get Reality right, you must first get Time right (as time is a key component of reality.)

Gravity and Holes

(Chapter 8)

In this chapter, we will talk about "Gravity," how science thinks it works, and what causes it to exist. Also, we will talk about the "Big Picture," and how gravity affects the way The Universe works.

Let's start with a simple definition of what gravity is (found on Google.) "Gravity, is a force of attraction between two particles, or bodies of mass." Seems simple enough. What causes gravity? The answer (also from Google,) "Gravitational effects are caused by the warping of spacetime." I guess this could be right? Never mind, something about this doesn't seem right! It looks like we are saying that "Gravity" causes things to move through spacetime, and that things moving through spacetime causes "Gravity." Also, it is said that gravity is neither energy or matter. So, does gravity move objects, or do moving objects create gravity?

In the I=(EM)$_r$ model of The Universe, gravity would have to be considered as information, which means that gravity Is some form of an Energy/Matter relationship.

I believe that "Gravity" is energy, created by matter, that affects matter. This would be an endless, (or almost endless) cycle, or it's a complete paradox, and we don't know anything. Science, by its own admission, does not fully understand gravity (just as it does not understand time.)

So, what started the endless cycle of gravity? To answer that, we need to go back to the "Big (outside looking in) Picture" of The Universe. From an inside the Universe view, understanding gravity is a huge deal. From the outside looking in view, it's an even bigger deal.

Newton's theory of gravity says, that everything in The Universe is pulling on everything else in The Universe. This is always an attractive force, always a pull, never a push. Newton's third law says, "For every action, there is an equal and opposite reaction."

This is where things get a little "Dicey." If gravity is always a pulling force, wouldn't there be an equal and opposite force as well? In "The Theory of Perpetuality," there is definitely an opposite force to gravity. Could this opposite force be called "Anti-Gravity?" Things that make you go "Hmmm."

Before we talk about anti-gravity, let's talk a little more about gravity, specifically large amounts of strong gravity. Let's talk about "Black Holes."

Black holes, are another subject that science has yet to completely understand, but does have a firm grasp on. This stands to reason, since we don't fully understand gravity. Not to mention, it's hard to study something you can't see (this is also the case with dark energy, dark matter and spacetime, as well as time.) However, we do know that they exist.

Black holes are the strongest known forces of gravity in the Universe. They can be found anywhere in The Universe. From the centers of galaxies, to the most remote corners of The Universe. Black holes come in a variety of sizes. Their sizes can range anywhere from ultra small, to super massive. We will probably never know how black holes work. If we were to cross the event horizon of a black hole, we would never be able to leave, or even send back any information learned.

In the Theory of Perpetuality, we are interested in something call a "Super Goliath Black Hole," that sits at the very end, and outside of The Universe. Before we talk more about the "SGBH," let's go back to that "equal and opposite," anti-gravity thing.

What is gravity's opposite? What is a black hole's opposite? These are important questions, as the answers may help us understand how The Universe works. We already know that anti-gravity is the opposite of gravity. Anti-gravity is almost never talked about, as a player in the how The Universe works game. As a matter of fact, according to General Relativity, anti-gravity cannot exist.

What we haven't talked about, is the opposite of a black hole, which is of course, a "White Hole." White holes are theoretical at this time.

White holes are described as:

- A reverse black hole.
- Repels energy and matter (information) away from it, with a force known as anti-gravity.
- Information can leave a white hole, but cannot be taken from it.

Sounds like the "Polar Opposite" of a black hole, which would produce the opposite effect of gravity, "Anti-Gravity."

In the Theory of Perpetuality, we are going to assume that white holes, and anti-gravity do exist (we are moving farther and farther from Newton's laws, and relativity by the minute.)

The gravity, anti-gravity relationship is an important one. This relationship has a lot to do with how The Universe begins, and how it ends! This relationship, may also have something to do with, how a current universe passes information on to a future universe!

We are going to talk more about black holes, white holes, gravity, and anti-gravity, in the chapters, "All Together Now," and "The Big Picture." Before we get to those chapters, there is one more hole to talk about. This would be the "Grey Hole!"

What is a Grey Hole? That's simple, a grey hole is any combined form of a black hole, and a white hole. A black hole and a white hole cannot be truly combined. A black hole cannot take the white hole's information, and the white hole would repel the black hole's information.

There is a scenario, where a grey hole could exist. This grey hole, would actually be a "Grey Hole Binary System." This system would consist of one black hole, and one white hole, who would orbit each other. There would be a polarized buffer zone between the black hole and the white hole, to prevent them from touching each other (where's the fun in that.)

Both the black hole, and the white hole would have their own accretion discs, but they would spin in opposite directions from each other. It is possible, that either or both the black hole, and white hole could be tidal locked.

There are four orbit variations, and four non-orbit variations to a "GHBS" (each variation includes a buffer zone.)

- White Hole Dominant (the white hole orbits the black hole, but is tidal locked.)
- Black hole dominant (the black hole orbits the white hole, but is tidal locked.)
- Alternating dominance (the black hole, and white hole alternate orbits, both are tidal locked.)
- Non-Dominant (the black hole and white hole always orbit each other, but neither is tidal locked. The black hole, and white hole are always orbiting each other, but the orbit speed is always fluctuating between them.)

- Non-Orbiting (the black hole and white hole sit side by side, but have their own rotation.
- Non-Orbiting, WHD (no orbit, white hole is tidal locked.)
- Non-Orbiting BHD (no orbit, black hole is tidal locked.)
- Null Orbit (the black hole and the white hole sit side by side; both are tidal locked.)

We know how black holes are formed, but how are white holes formed? The answer could be, a white hole is actually an inverted black hole. I'm not going to get into all of that, as it may require a book of its own.

There is much that we don't know about Black holes and White holes, gravity and anti-gravity, but we will get there.

Thought

Anything caught in the buffer zone of a grey hole, might be crushed into a singularity of its own. This new type of singularity would be called a "David Zone Hole," and could be a portal to other "Realms!"

Thought

It is possible that black holes have a network connection with other black holes. The same is not true of white holes.

Thought

Just think, gravity all started when an apple fell on some guy's head!

Now, let's get to the "Big Picture" purpose of black holes, white holes, and grey holes (we're just talking about the really big ones.) These "really big ones" are the Super Goliath Black Hole, and Super Goliath White Hole, or if these Super Goliath's would combine, they would form the Grey Hole Binary System! We could call this the *Super Massive Goliath Binary Grey Hole System*, but that seems like a bit much (it's too much magic, too much magic.)

We can now go through the process of the beginning and ending of The Universe, using both the Goliath model and the Grey Hole model.

Let's start with the "Goliath" model. In the beginning, all possible information entered The Universe. But what and where did it enter? Good question, information entered The Universe through the "Beginning Portal." This portal is known as the Super Goliath White Hole, or SGWH. The "Gifted" information entered the SGWH, and was immediately repelled (by anti-gravity,) into the void. Once in the void, information was attached to time, and spacetime (and reality) were born. This was the beginning of the process we call "Inflation."

At the other end of The Universe, lies the Super Goliath Black Hole, or SGBH. The SGBH, using its gravity, is what is responsible for "Expansion." Inflation and expansion (anti-gravity and gravity) has, or had an information control ratio (kind of a push/pull ratio.) This ratio started out being heavy on the inflation side. Soon the ratio would begin to balance out, creating a drifting or coasting effect. Near the midpoint of The Universes existence, expansion and gravity would take more control.

The expansion, or velocity of information through The Universe, would seem to be on the increase.

At the end of The Universe, gravity will have pulled everything into the SGBH. This SGBH, as well as spacetime and its historical information record, will then collapse. This collapsed Universe will then invert, becoming the next universes SGWH. In this "Goliath" model, each universe will pass on its existence (upon its death) to the next universe.

The process is the same in the "Grey Hole Binary System" model of The Universe. There are two differences. The first difference is, the GHBS is the beginning and the end. The reason for that, is the second difference. The second difference is, in the "Goliath" model, The Universe is a "Flat Universe." In the "Binary" model, The Universe is "Folded!"

There's only one more thing to talk about in this chapter. Where does "Gravity/Anti-Gravity originally come from. This one might surprise you! You may be completely "Wowed," or you may think, "This guy's a complete idiot!" Anyhow, here goes.

Before "The Beginning," there was nothing but "Time." This "Time," was unburdened (meaning it had no information attached to it, and was not a part of any reality.) When time ran into "The Void," and all of this new information was attached to it, "Time" became burdened.

This is where things get interesting. It turns out that "Time" has what is called, "Temporal Velocity," also known as the speed of time! "Unburdened Time" and "Burdened Time" travel at completely different temporal velocities.

The temporal velocity of unburdened time, is something like the speed of light squared, or $(SoL)^2$. This cannot be proven, just as nothing about unburdened time can be proven, because it does not exist in our reality. The temporal velocity of burdened time is much, much slower. The velocity of burdened time, can depend on the amount of information that attaches to it, and the amount of matter/mass, and the types of energy that are attached to it.

Why does any of this matter? When unburdened time slams into the new universe, it's like a speeding car running into a parked car. At the point of impact, the parked car explodes forward, as the speeding car hits it. The velocity of the speeding car slows comparatively to a crawl.

Since time is omnipresent, the portion of time that hits The Universe, and becomes burdened, is relatively small. The larger, unburdened portion of time, continues, blowing past The Universe, and its newly burdened time. The effect that time, in its unburdened form, has on The Universe, in its passing of The Universe, is where gravity originates from. It's like a truck blowing past you on the highway. You can feel its pulling effect. Just as in the case of unburdened time causing gravity, unburdened time also causes, or is responsible for creating anti-gravity. That temporal crash of "Time" into The Universe, creates that pushing, anti-gravity effect.

This leads to the first of three laws of Temporal Velocity. "Temporal Velocity can create physical velocity." The second law of temporal velocity states, "Temporal Velocity can create physical force." Finally, the third law of temporal velocity says, "Time can travel at unburdened temporal velocity, and burdened temporal velocity at the same time, while still remaining connected."

Time traveling at multiple temporal velocities at the same time, creates what is known as "Sling Time." It's like pulling a rubber band back, then letting it go. Eventually, time traveling at the slower, burdened temporal velocity, will rejoin the faster moving portion of itself, and time will remain constant. The temporal velocity of "Sling Time" is variable. It will start out at its current burdened temporal velocity, then increase well beyond unburdened temporal velocity, until it returns to current unburdened time.

It looks as if "Time" is a whole lot more than a measurement!

*****all other theories are crap! *****

Between Chapter "Crap"

List of Possible Dimensions

- Physical
    - Length
    - Hight
    - Width
- Temporal
    - Burdened
        - Past
        - Present
    - Unburdened
- Reality (this includes past, present, possibilities and impossibilities.)
    - Burdened
    - Unburdened
- Universal – Realities Shared by Multiple Universes.
- Realm – Identical Universes that exist in more than one realm.
- Omnipotent – A Dimension (or super realm) where all Realms exist as one.

NOTE

Since time is omnipresent, it exists in every dimension.

"All Together Now!"

Chapter 9

In this chapter, we will spell it out, step by step, from beginning to end. The model of The Universe used here, is an original, folded, grey hole universe, with a "GO" creation, and an undetermined ending. We will not talk about TARs in this exorcise.

After spelling it out, we will try to see if this is a valid model of how The Universe Works. Let's begin!

- Before the beginning, only infinite, unburdened time exists.
- The void that would become spacetime is created.
- All possible, "Gifted" information, is prepared to enter the void, via the White Hole Dominant, Grey Hole (WHDGH.)
- All possible information, is pushed through the WHDGH, into the void by time, converting the void into space.
- Time enters the void, with all information, which becomes attached to time. Time is converted from its original unburdened form, into its much slower, burdened form.
- Burdened time and space become one, to form spacetime.

- The initial collision of unburdened time, with the WHDGH, creates the effects of anti-gravity. The "GO" process of inflation begins.
- Possibilities begin to become actuality. Actualities exist in the form of normal energy, and normal matter.
- As a result of the new actualities, some possible information becomes impossibilities. Impossibilities exist in the form of dark energy, and dark matter.
- Actualities immediately become actualities past.
  - Information in the grey past, exists as grey negative energy, and grey antimatter. In this situation, the term "grey," refers to the ability of portions of the negative energy, and antimatter, to be seen in the present, as well as existing in the past.
  - Information in the dark past, exist as negative energy, and antimatter.
- As reality plays out, spacetime records the history of reality.
- Anti-gravitational inflation begins to subside, as the WHDGH begins to transform into a Black Hole Dominant, Grey Hole (BHDGH.)
- Gravity from the BHDGH, and any anti-gravity that remains from the WHDGH, begin the process of expansion.

- The process of information, shifting into different phases of reality, continues until all remaining possibilities have been depleted.
- Only impossibilities remain (in the form of dark energy, and dark matter.) Essentially, The Universe, and reality will be dead! There will be no grey past, and the dark past will be completely historical.
- The Universes final act, will be for spacetime, and its historical record of reality, to collapse.
- The collapsed historical record of reality (which includes a copy of the original "all possible information,") will be pulled into the "David" buffer zone, that exists between the now nondominant black hole, and white hole that make up the Binary Grey Hole System (BGHS).
- Time (inside the David zone) will become unburdened for one SMAT, before pushing all possible information into the next universe.
- The process ends, and begins again!

Now for the fun part. Can any of this be proven? More importantly, can any of this be disproved? The following chart will attempt to answer these questions.

	Description	Can Be Proved	Can Not Be Proved Can Not Be Disproved	Can Be Disproved
1	Unburdened Time		√	
2	Burdened Time	√		
3	Time is constant		√	
4	Time is Infinite		√	
5	All Possible Information at "GO"	√		
6	Information "Gifted" by GOD		√	
7	Universe Begins as a Void		√	
8	Grey Hole Binary System as Beginning Portal		√	
9	I = (EM)r	√		
10	Information Attaches to Time	√		
11	Catagories of Information	√		
12	Phases of Reality	√		
13	Energy and Matter Types and Their Functions		√	
14	Void Becomes Spacetime	√		
15	Sub-Spacetime Records All Historical Information		√	
16	Temporal Impact Causes Anti-Gravity		√	
17	Unburdened Temporal Drift Causes Gravity		√	
18	Inflation Caused by Temporal Anti-Gravity		√	
19	Expansion Caused by Temporal Gravity		√	
20	Grey Hole Binary System as Ending Portal		√	
21	The Present is 1 SMAT	√		
22	There Are No Alternate Universes		√	
23	There Are Temporary Alternate Realities		√	
24	Normal Energy	√		
25	Normal Matter	√		
26	Negative Energy	√		
27	Antimatter	√		
28	Dark Energy	√		
29	Dark Matter	√		
30	Grey Energy		√	
31	Grey Matter		√	
32	Negative Grey Energy		√	
33	Negative Grey Matter		√	
34	Negative Dark Energy		√	
35	Negative Dark Matter		√	
36	Possibilities End	√		
37	Reality Ends	√		
38	Spacetime Colapses		√	
39	Historical Spacetime Singularity is Created		√	
40	"David Zone"		√	
41	Historical Spacetime Singularity Creates New Universe		√	
42	Multiverse System Exists		√	
43	Omniverse System Exists		√	
44	Time Remains Unburdened, Infinite and Constant		√	
45	Burdened Temporal Velocity		√	
46	Unburdened Temporal Velocity		√	
47	Temporal Adhesion	√		
48	There is no Actual Future	√		
49	Time Travel		√	
50	***All Other Theories Are Crap!"***		√	

It appears as if the "Theory of Perpetuality" may be a viable theory and is not *"complete crap!"* Notice that the "Can Be Disproved" column has no checkmarks. It is however just a theory!

*** *all other theories are crap!* ***

The Theory of Perpetuality

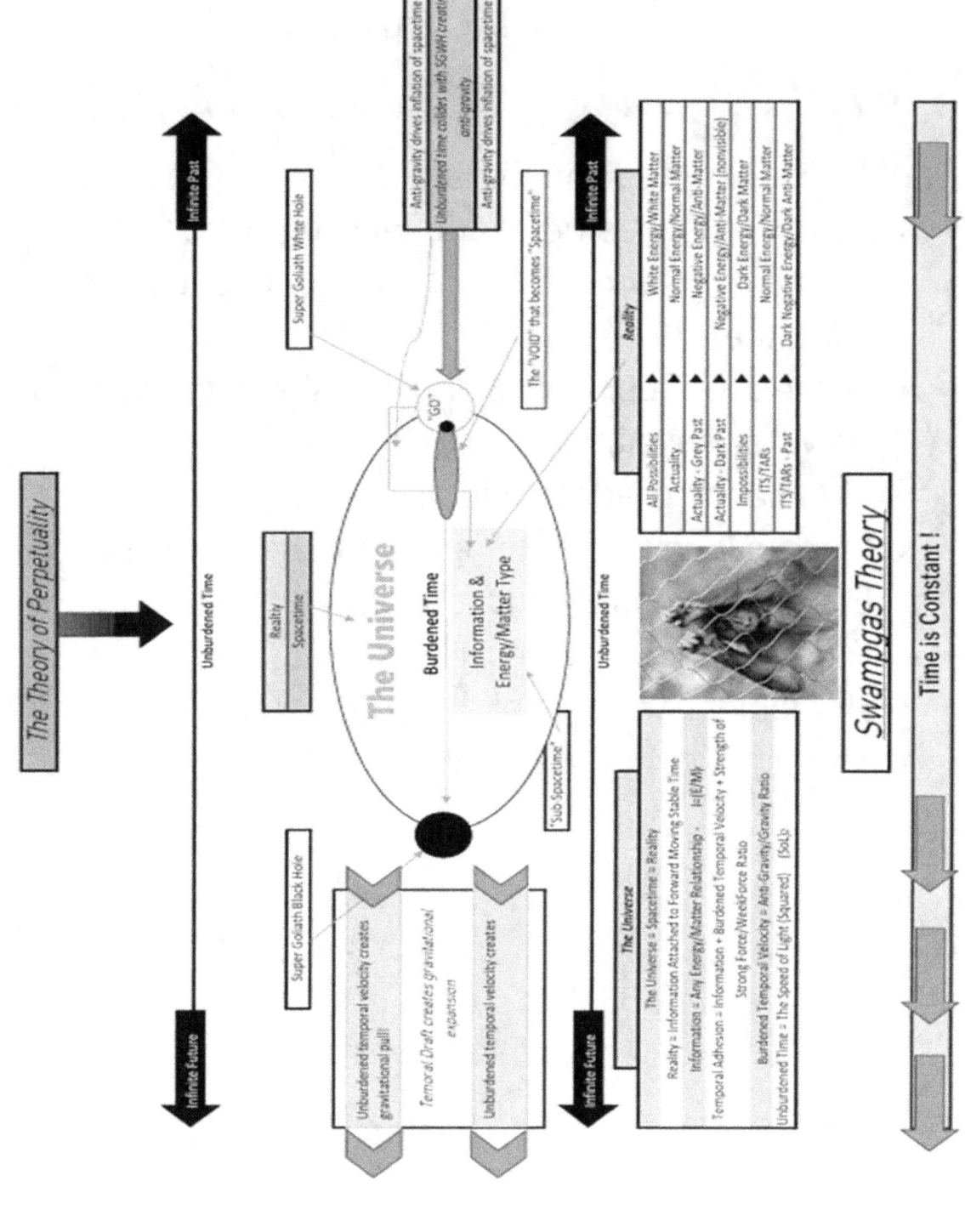

Infinite Past

Super Goliath White Hole

Anti-gravity drives inflation of spacetime
Unburdened time collides with SGWH creating anti-gravity
Anti-gravity drives inflation of spacetime

The "VOID" that becomes "Spacetime"

Infinite Past

Unburdened Time

Reality
Spacetime

"GD"

The Universe

Burdened Time

Information & Energy/Matter Type

"Sub Spacetime"

Unburdened Time

Reality

All Possibilities	White Energy/White Matter
Actuality	Normal Energy/Normal Matter
Actuality - Grey Past	Negative Energy/Anti-Matter
Actuality - Dark Past	Negative Energy/Anti-Matter (nonvisible)
Impossibilities	Dark Energy/Dark Matter
ITs/TARs	Normal Energy/Normal Matter
ITs/TARs - Past	Dark Negative Energy/Dark Anti-Matter

Super Goliath Black Hole

Unburdened temporal velocity creates gravitational pull

Temporal Draft creates gravitational expansion

Unburdened temporal velocity creates

Infinite Future

Infinite Future

The Universe

The Universe = Spacetime = Reality
Reality = Information Attached to Forward Moving Stable Time
Information = Any Energy/Matter Relationship - I=(E/M)
Temporal Adhesion = Information + Burdened Temporal Velocity + Strength of Strong Force/Week Force Ratio
Burdened Temporal Velocity = Anti-Gravity/Gravity Ratio
Unburdened Time = The Speed of Light (Squared) (SOLb)

Swampgas Theory

Time is Constant !

Between Chapter "Crap"

Universal Truth's

- Time is Infinite and Constant.
- Alien life is, has been and will remain on Earth.
- Aliens have altered "Mankind" genetically. This will continue.
- There are "Supreme Beings" that exist outside of our universe's realm.
- Man has degraded religion (badly.)
- We know next to nothing about The Universe.
- Everything has a polar opposite.
- Closed minds prevent true knowledge.
- Humanity is one race! It should not be a divided one.
- The Truth is full of lies.

"The Big Picture"

Chapter 10

In this chapter, we will look at the different types of universes. We will also discuss when universes group together to become a multiverse. Finally, we will talk about the omniverse.

There are four categories of universe types.

- How they receive their possibilities.
- How they begin and end (portal types.)
- The shape of the universe.
- What happens when they die.

Let's start with "Category One." How universes receive their possibilities.

- "Gifted." All possible information, is gifted to its universe by a higher power. Only an "Original" universe, has gifted information.
- "Refreshed." All possible information comes from the previous universe. A copy of the previous universes recorded historical information is refreshed (the history of the information has been removed.) This "Refreshed" information then enters the new universe, via its beginning portal.
- "Resurrected." A resurrected universe, uses its refreshed historical information to become the "All

possible information," in its resurrected form. The ending portal of the dying universe, becomes the beginning portal of the newly resurrected universe.

Now let's look at "Category Two." Beginning and ending portal types. There are four portal types that can exist as a part of any universe.

- Beginning portal, is a "Goliath White Hole," and the ending portal is a "Goliath Black Hole."
- Both the beginning, and the ending portals are separate "Binary Grey Hole Systems."
- The beginning, and the ending portals are the same "Binary Grey Hole System."
- Beginning portal, is a "Goliath White Hole." There is no ending portal (this type of universe does not replicate its historical information. When it dies, that's the end.)

Next, we have "Category Three." The type of shapes that a universe can have.

- "Flat." A flat Universe, has a beginning at one end of the universe, and an ending at the other.
- "Spherical." This type of universe, could be round, or nearly round. In a spherical Universe, you cannot always see its beginning.

- "Folded." A folded universe, would have a single GHBS as both its beginning, and its end. Inflation would push to its farthest reaches, and expansion would bring it back home.

Now let's check out "Category Four." What happens when a universe dies. Some of these are the same, as some of the beginning types.

- "Refreshing." All possible information, comes from the previous universe. A copy of the previous universes recorded historical information is refreshed (the history of the information has been removed.) This "Refreshed" information then enters the new universe, via its beginning portal.
- "Resurrecting." A resurrected universe, uses its refreshed historical information, to become the "All possible information" in its resurrected form. The ending portal of the dying universe, becomes the beginning portal of the newly resurrected universe.
- "Non-Procreational." An NP universe type, does not pass on its information. When it's over, it's over.

Universes can take on many forms from these four categories. I believe our universe to be a "Gifted: or original, BGHS, Folded, Resurrecting universe. It is possible, that there could be something called a

"Baseline" universe. A baseline universe, is one which has no free will. Its purpose is to optimize, and carry out its existence, as the best outcome possible. This could also be an experimental universe.

Now that we've talked about the types of universes, let's move on to "The Multiverse." There are three types of multiverses.

- Concurrent Multiverse. Each universe in this type of multiverse, exists one at a time. Each universe receives its information from the previous universe.
- Congruent Multiverse. Universes in this type of multiverse, exist all at the same time. They receive their information, from the original universes collapsed historical spacetime singularity.
- Overlapping, or Bubble Multiverse. The universes in this type of multiverse, can be concurrent, or congruent. They can overlap, and interact with each other. Their information is originally gifted, but can also be easily shared through an elaborate grey hole network.

Note

There can be some "Rogue Universes" out there, that do not belong to any multiverse. These are rare, and do not function as well as universes that belong to a multiverse.

Finally, we'll define what an "Omniverse" is. This one is simple. The omniverse, and there is only one of these, is a collection of all of the multiverses, along with any rogue universes that exist, or have ever existed. When the Omniverse is over, it really is, "Game Over."

At this point, all that will exist is time. The form of time that will exist, will be entirely new. This form of time, is called "Historically Burdened Time."

*** *all other theories are crap!* ***

Between Chapter "Crap"

"Spacetime Voids"

Is it possible, that there is a place, or there are places, where time and space do not come together, to form spacetime? As far as I know, there is no evidence of such a place. Given the small amount of The Universe that we know anything about, I believe that it is definitely, a possibility.

If there is such a place, its purpose may be for the existence of wormholes, or even having something to do with quantum entanglement. It may also be a place, where burdened time may temporarily become unburdened, for reasons unknown.

There is also the possibility, that any spacetime voids, may lend some flexibility to spacetime, to ensure that there is a bend/don't break failsafe, in place for the fabric of time.

Another scenario, could be that spacetime voids, could be a linking point, between universes that exist in a multiverse. There may be no apparent reasons for the existence of spacetime voids

Is it possible that they exist? Yes, it is.

"Free Will"

Chapter 11

Is there freewill in The Universe? The answer to this question seems to change, depending on where you look.

Let's take a look at what science has to say. In quantum physics, the answer started out as a "hard no." Later, science allowed some room for free will, but that didn't fit in with relativity. At that point there was an attempt to change the definition of free will. So, let's just say that science is all over the place on this one!

By the way, science also says that every time a decision is made, or a possibility is created, a new universe is formed. That sounds like a "hard yes" on free will to me (it also sounds like a load of crap.) Like I said, "all over the place."

Since science isn't very helpful on this question, where else can we look? We can look at what government and society says. In most societies, we are rewarded for making good decisions, and punished for making bad ones. This would be an indication of a belief in free will.

We can look for answers in another place, "Religion." It feels like (in general) religion falls on the "yes" side. But after a closer look, religion might go either way, depending on its benefit to them. Being born a sinner, feels like predetermination. Being able to choose atonement for those sins, feels like free will. Religion allows for judgements based on choices made. Religion is confusing. Science and society, are also confusing.

Now, for one more look, let's go to philosophy. In philosophy, there are many definitions for free will. These definitions imply that philosophy lands in the "yes" column.

Is there free will in The Universe? The best place to find the answer to this question, may be inside ourselves. What does logic tell you? What does your heart tell you? What does your soul tell you? The answer for me, is always Yes.

If the outcome of The Universe is predestined, then what would the point of The Universe be? It may be some sort of a science experiment for a supreme race of beings. It could be a form of entertainment, or even some kind of a game. Perhaps The Universe is nothing more than a single thought from one individual, belonging to an omnipotent race of beings.

But for me, a "no" answer would mean that life has no purpose, and no meaning, and I know that this is not the case!

Since this book is based (loosely) on science, and science treats relativity and quantum physics as marriage partners, (not happily, possibly an arranged marriage) we may need to do what married couples do when they don't get along. Say "whatever" and live to fight another day. Or, maybe we need to find a compromise that works.

If you're going to prove free will mathematically, you need to start with the correct information. Otherwise, it's "garbage in, garbage out." What am I saying here? What I'm saying is that some of the information, in either or both relativity and quantum mechanics, could be garbage. If we are going to prove whether free will exists, we will first have to sort through our garbage.

Let's say that science eventually goes the "yes" route. There may be some factors to consider, in making it make sense.

There are many stages to "Free Will."

- Complete – To create or act by thought alone. this is limited to "Supreme" sentient beings.
- Realized – Choice is realized by thought, but implemented by physical action. This type of free will, is held by "Sentient," non omnipotent beings. There are four types of realized free will.
  - Reasoned – These are well thought out choices, that coincide with a desired outcome. These free will choices have intentions (good or bad.)

- Indifferent – Pick one, or flip a coin. The outcome doesn't seem to matter, one way or the other.
- Indecisive – Pick one, unable to determine the best choice
- Undecisive – No choice is made. A lack of caring or interest in making the best choice.

- Unrealized – Autonomic thought, with physical action. This type of free will, can be held by sentient, as well as non-sentient life forms.
- Non-Existent (Null) – This is the case, in non-optimally operating life forms.

At this point. I need to re-ask myself, what is free will? The answer is: The ability to make a choice, and then to act upon that choice (making no choice, is the same as making a choice.)

As far as science goes, we need to get this answer right, if we want a "clear truth" picture of The Universe.

As far as "Me, personally," The answer remains a definite "Yes."

As far as The Universe goes, I believe that The Universe has "limited free will." That's a bold statement. This would mean that The Universe, is some sort of a life form. A life form that possesses intelligence.

Not only does The Universe have free will, its free will supersedes our own free will. If need be, The Universe has the ability to persuade our choices, or even change the choices that are available to us. This would be extremely rare, and would only take place if an undesirable event were to occur at the expense of The Universe.

The Universe could use an ITS, to determine the probability of a negative outcome, and if need be, a TAR, to be used as a course correction on the negative event. Any (universal) catastrophic events caused by free will choices, will not be allowed. Universal free will, always trumps personal free will. These situations would be very rare, and hopefully never have to occur.

In conclusion, I'm going to say, that free will does exist in The Universe, and by The Universe. "Who the hell is Will anyway? and why does he always require our deliverance?"

*** *all other theories are crap!* ***

Between Chapter "Crap"

"Swampgas Mass"

There are a couple of things, that are technically an energy/matter relationship, whose mass cannot be measured (or at least not by us.) The first of which, is light. The second of which, is thought.

Let's talk about light first. Light is defined as energy that does not occupy space, and has no mass; therefore, it cannot be considered matter. There are a couple of problems with this definition. The way we define mass, is through atomic consistency (neutrons, protons, and all that.)

Everything in The Universe is considered to be information. This information is always an energy/matter relationship. Information is measured by its E/M ratio. All E/M ratios, must have some amount of energy, and some amount of matter (mass.) The amount of energy, or mass, can be so small, that we cannot measure it. This does not mean that it does not exist.

To say that light takes up no space, and has no mass, is incorrect. If you can see it, it has mass, and it takes up space. If you can feel it, it has mass. Everything in The Universe has mass, weather you can measure it or not. Matter (which has mass) is a byproduct of energy, and energy, is a byproduct of matter. All forms of information, have both energy, and matter/mass. We have not come up with a way to measure light's mass. That does not mean than light has no mass. In this book, unmeasurable mass is now known as, "Swampgas mass."

What about thought, does it have mass? Since thought is a form off information, it must have mass. Thought may have an entirely different type of mass. Thought's mass, is what we call "Associative mass." This type of mass is a very small (nearly unmeasurable) percentage all of the different matter, that is involved in creating said thought. In this situation, "Swampgas ass. mass," is not measurable.

It's something to think about.

"Eighty-Eight Miles Per Hour"

Chapter 12

I probably don't need to write an introduction for this chapter. Clearly, it's about time travel (finally, something fun to talk about, or at least fantasize about.) There are many Sci-Fi theories on time travel, and how time travel would work.

But what about actual science-based theories? There are some theories about time travel out there, most of which are based on relativity. Relativity says that time can speed up, or slow down, based upon certain conditions. This, however may not be entirely true!

Relativity does not understand what time is, or its role in the reality of our universe. Before we look at any "Swampgas" theories on time travel, let's review what we know about time.

- Time is constant.
- Time is infinite.
- Time is burdened (inside The Universe.)
- Time is unburdened (outside of The Universe.)

Time is constant. This means that time does not speed up, or slow down, for any reason, (sorry relativity.)

What can be changed, is information's temporal adhesion to time. By changing the strength of information's temporal adhesion to time, that information can actually travel faster, or slower through time, than any information that is moving at the "Standard Temporal Velocity."

Time is infinite. This is a tough one to explain. It's more than the fact that time has always existed, and more than the fact that it always will exist. Every moment (or SMAT) in time that has ever existed, will continue to exist, as part of times infinite timeline. This existence of time's past, is not just historical, it is physical. Every past moment consists of real information. That means, there is real energy, and real matter in our universe, that makes up the entire past existence of our reality. Wow! Time is also omnipresent. This means that time is everywhere, all of the time, at the same time.

Time is burdened in our universe. This means that time has information attached to it. Also, the standard temporal velocity of burdened time, is much slower than the temporal velocity of unburdened time. When we talk about time travel, we are talking about traveling through burdened time.

Time is unburdened outside of our universe. Time travel is not possible through unburdened time, since there is no reality associated with unburdened time.

Now that we've reviewed what time is, there are a few questions that need to be asked about time travel.

- Is time travel possible?

- Is time travel to the future possible?
- Is time travel to the past possible?
- What (if any) are the rules of time travel?
- If time travel is possible, how do we achieve it?

Is time travel possible? Of course it is. We do it each and every day, one moment (or one SMAT) at a time. We travel through time, at the standard, burdened temporal velocity (due to our temporal adhesion to time.) We move through time, minute by minute, day by day, year by year, and so on, throughout our entire lives.

What we really want to know is, can we alter our temporal velocity to achieve time travel? Unfortunately, we may never know the answer to this question.

Is traveling to the future possible? Since the future has not been written, it is not possible to travel to a definite future. What may be possible, is traveling to any one of a very large number of probable futures. If you want to see what your future might look like, you would want to choose a possible future, whose probability of becoming reality, is as close to 100% as you can get. This is where ITS' (if then scenarios) come into play.

The chance of landing in an accurate possible future, is very slim. The farther you jump into the future, the chances of it being accurate, are greatly reduced.

Let's assume that traveling to a possible future is possible. How would we achieve this. To achieve future time travel, we would have to adjust our temporal adhesion, which would in turn, change our temporal velocity. A good place to start, would be understanding temporal adhesion's function.

Temporal adhesion, is what keeps information always in the present, and keeps the present stable. The actual temporal velocity of information in our reality, is dependent on the strength of that information's temporal adhesion. In normal reality, the standard temporal velocity, is achieved by having a temporal adhesion strength that is continuously stable (always the same.)

It's because of the adhesion strength of standard temporal velocity, (which is set somewhere in the middle of the scale,) that we are always moving through time slower than the velocity of (burdened) time itself. This may be hard to comprehend, but it is very true. In effect, our present, is in the burdened temporal past.

Back to traveling in time to the future. If we want to jump to the future (or a possible future,) we would need to increase the strength of our temporal adhesion. Basically, we would tighten the grip of our attachment to time. By doing so, we would be moving faster in time, than any information that is traveling at the standard temporal velocity.

How do we do that? To strengthen or loosen our grip on time, we would have to change the weak force/ strong force ratio, that affects the information that is attempting to time travel. Don't ask me how to do this, because I really don't know.

So, by adjusting our adhesion strength to time, we can reach any desired future. When we do reach that future, we would have to return to the standard temporal velocity, to prevent us from bypassing our destination.

If you do travel to a possible future, you cannot return to your original jump point. You can however, loosen your temporal adhesion to any point below standard temporal velocity. This would allow reality to catch up with you. The looser the adhesion strength, the faster reality catches up.

The reason you cannot jump to an actual future, is the same reason that the future has not been written. That reason is this, there is no such thing as "The Future." Here is how the past, present, future thing works.

Let's start with the present. This is where we always are. The present immediately becomes the past. As the present becomes the past, we move on to a new present. The past gets larger as time moves on. The present is only a single moment in time. Present becomes past, present becomes past, etc... This happens continually throughout time. There is no future, only possible futures.

To summarize, future time travel, is not very probable. If it were, there would definitely be some rules that would apply.

Since time travel to the future might be too difficult to conquer, can we at least travel to the past? It might be possible, but there is much more to overcome. There are many rules to past time travel. Also, the mechanics of going back in time, is very complicated.

Going back in time, will require more than a flux capacitor and a DeLorean. We will need to find a way to become totally detached from time, while still remaining inside of our universe. Temporal rift? Wormhole? Who knows. But we will have to figure this out to make it work. Once we reach our destination, we will have to reattach to time. This brings us to question four.

Are there rules to time travel? The answer is yes, absolutely, there are many rules.

1. Time travel cannot, and must not, affect any actualities timelines, past, present, or possible future.
2. To ensure the enforcement of rule one, all time travel is "view only." You can see everything that is happening. Your existence will not be known by your surroundings. There can be no interaction between you and the timeline you are visiting.

3. When traveling to the past, you cannot visit a timeline where you already exist. This includes anytime from your conception to the present. Landing at a time where you already exist, would cause a "Duality Paradox," and would not be allowed.
4. When traveling to a possible future, you cannot land at a time where you could possibly exist. This would greatly reduce the probability of it becoming an actuality for your possible future.
5. When traveling to a possible future, you cannot select a future where you didn't exist in that future's past.
6. The Universe can institute the use of TARs to protect rule one. The use of a TAR will ensure that your absence will not be known by your surroundings, when you make a "Time Jump." No one will ever know you were gone (including you.)
7. When returning from a time jump, you can only return to the "Sync Point" of the TAR and the original timeline. An example would be, if you were gone for one year, you would have to return to your timeline one year after your time jump began.
8. Memories (of any events) gained from time travel, cannot return with the time traveler to the present timeline.
9. You cannot jump to a possible future that is beyond the "current burdened present." Remember, we exist in burdened time's past.

With the almost unachievable technology needed for time travel, and the uselessness that time travel would bring (due to the rules of time travel,) there may not be much point in spending a lot of "Time" on time travel. This is a very disappointing conclusion. I guess we can still fantasize about time travel. To make time travel worth the effort, we may have to figure out a way to break some of the (time travel) rules.

****Notes****

- Traveling through a wormhole would require some serious gravitational shielding.
- Time Travel is only possible, in a universe where free will exists. Quantum Determinism does not allow for time travel.
- Time Jumps can be "objectionably bound" (your reality jumps with you) or "temporally bound" (you will jump without your surroundings.)

"The time is gone; the song is over. Thought I'd something more to say. (Pink Floyd)

***all other theories are crap! ***

Between Chapter "Crap"

"The Shape & Speed of Time"

The shape of time, can be broken down into two parts.

- True Time, or Unburdened Time, has no determinable shape. It is everywhere, all of the time, at the same time.
- Real time, or Burdened time, takes on the shape of the information that is attached to it, and where the attached information is located in The Universe. This allows for time staying constant, in a folded universe. This is one of the reasons, why Einstein believed that time could bend, or be bent. This is actually an illusion, caused by temporal velocity, and temporal adhesion.

The Speed of Time, can also be broken down into two parts.

- The speed of True Time is $(SoL)^2$
- The Speed of Real Time is $(SoL)^2$ x wf/sf

*** The speed of time has no shape and the shape of time has no speed. ***

Part Two – Swampgas Theories

(Some other Out of This World Theories)

***these theories could be crap! ***

"Heaven and Hell"

Chapter 13

Are Heaven and Hell real? For many, the answer is yes. Yes, because most religions tell us so. For many others, the answer is no. No, because there is no actual proof that they exist.

The "No's," believe that heaven and hell, are manmade concepts. These concepts, are designed with promises of great rewards, for good behavior, and following the guidelines of their religion, and threaten bad behavior, with the punishment of great terror and pain.

The "Yes'," have a combination of belief, and a fear of non-belief, in their religion (I believe because the consequences of nonbelief are too great.) The fear of not believing, is not the same as believing. This is why I believe that true faith cannot be attained. There will always be a percentage of doubt. Just as true faith is not possible, true nonbelief is also not possible. Non-believers, will always question whether their choice is the correct one. There is a chance, that they could be wrong (and that's scary.) Faith in the unknown can be hard. Both faith and proof, will always come with doubts.

Let's go back to the original question. Are Heaven and Hell real? The answer is, we just don't know. It is possible, that there are "Realms" that could exist, whose descriptions resemble those of heaven and hell.

We'll start with the realm of heaven. This realm is a "Free Time" realm. Both time, and all possibilities, live here. Information in this realm is not attached to time. There is no organized reality here (it's only what you make of it, or what you want it to be.) It is a true "Nexus" (in the Start Trek sense of the word.)

This is a place, where you can experience anything you want, anytime you want, and as many times as you want. Everything is always possible. Because information is not attached to time, you can re-live, or create new reality events whenever you want to.

Happiness and serenity live here. This realm is not limited to just the souls of humans. All life forms can exist here (including advanced life.) This could be the realm where all possible information comes from.

Now, let's look at the realm of "Hell." This is a "Timebound" realm. Time, and all possibilities exist here. Time is "not" constant here. Because of time existing in a skewed form, information can attach to time at any point in time, as many times as it wants. All possible information, can exist at the same point in time, all of the time. This loop can go on endlessly.

There is no past, and no possible future. The timebound realm, is in constant chaos. Gloom, misery, and fear live here. This is a world of nightmares. Avoid this realm at all costs.

It is possible, that heaven and hell, are both in the same realm. Your ability to navigate in this realm, will determine your experience.

There may be a third realm, that is neither heaven or hell. This is the realm of "True Death." This is known as a "Dead Time" realm. This realm, is the easiest to describe, but the scariest to think about. There are two possible versions of this realm.

- Version one is simple, all possibilities exist, but you do not.
- In version two (which is in my opinion, the worst of the two) all possibilities exist. You also exist in version two. You have knowledge that all possibilities exist, but you have no access to them. In this realm, you are forever alone (in darkness.) This may be worse than actual death.

The final possibility of realms, that could be heaven or hell, is the lack of possibilities. Without possibilities, "Death," really is "The End." The only thing worse than having not existed, is existing with the knowledge of having that existence, permanently end! This type of existence can feel pointless, and can lead to a desensitized value of life. This is one of the hardest things for humans to think about, or talk about.

Personally, I believe that death is not the end. Our journey has only just begun. Throughout time, knowledge and wonder awaits us!

The topic of what happens after death, whether it be heaven or hell, or any other great unknown, is a major source of anxiety, panic attacks, and stress, for many people. Talking about the causes of anxiety, is a great tool in overcoming it!

*** *this theory could be crap!* ***

Between Chapter "Crap"

"That scene in Waynes World!"

The one where Garth, makes eye contact with his dream lady (AKA "Foxy Lady.") He and his chair, are thrown backwards into a metal pole, at the speed of light. With the sound of a clank, Garth gets up, lovestruck and dumbfounded, he returns to his table at the donut shop.

This is an example of something good, causing anxiety. Anxiety, at times, can be a good thing. We tend to dwell on anxiety, as being a bad thing (most of the time it is.)

Through the years, I've learned to control my anxiety (both good and bad.) As it turns out, I wasn't really controlling the bad. I was suppressing it, hiding from it, ignoring it, and hoping it would just go away. Hiding from the causes of anxiety, only leads to a bigger problem down the road.

For me, one of the biggest causes of anxiety, was death (and jury duty.) Not so much the thought of dying, but the thought of no longer existing. Time, and infinity were a very scary thing. The thought of me not existing, in an infinite past, didn't bother me. The thought of me not existing, in an infinite future, now that was a problem (and still is.) I am definitely not ok with that. The thought of this, still throws me (and my chair) backward, (at the speed of light,) into a metal pole. Every time!

This is what lead me to write this book. Figuring out for myself, how things work, has helped me face my anxiety, and its causes. Talking about anxiety, is a much better approach than hiding from it.

The importance of talking about mental health, should not be minimized. Sharing with others, as well as listening to others, can be a great tool in achieving good mental health. Remember, anticipation is always worse than the actual event!

Party on Wayne, and party on Garth!

"Sentient Life"

Chapter 14

Most define sentient life, as any life that has the ability to have feelings, to experience sensations and emotions. I believe there is more to it than that. You must be self-aware, and be affected by the knowledge, that you are self-aware. You must also, have the ability to grow through knowledge. But there is still more to it than that. To be fully sentient, you must have a body, and a soul.

Let's take a look at the first ingredient needed to be considered a sentient life form, "The Body." The body is what makes us physically alive. A better definition of the body, is Information, gifted by a higher power, or a supreme being, which through the natural unfolding of The Universe, becomes life. However, a body without a soul, is just an empty shell. A house, but not a home.

Now, let's look at the second ingredient, the soul. But what is a soul? There are many words used to define what a soul is.

- Immaterial Essence
- The Mind
- Emotions
- Character
- Thought

- Feelings
- Energy
- Consciousness
- Personality

These words, do not do the definition of a soul justice. This list shows the difficulty, in explaining, and in understanding what a soul is.

Just as the information that makes up our bodies is gifted, the soul is also gifted. The difference between our gifted bodies, and our gifted souls, is this. Our bodies are made of information that exists in our universe, souls are gifted from another realm altogether, and (pre-arrival) have nothing to do with our universe.

The soul's purpose, is to experience life, from the perspective of being in our universe, and some day, returning to its own realm, to share those experiences with others. There is no scientific proof of a soul. There is no religious explanation of a soul, that does it justice. Also, there is no adequate philosophical concept, of what a soul is.

The only way a soul, can experience life outside of its own realm, is to experience it while attached to a life form from inside our universe (that would be us.)

This is where things become a little clearer, at least they do for me. Knowing now, that a sentient being requires a body and a soul, we can define and discuss, what the body/soul relationship is, and how this relationship works.

A new and better definition of sentient life, is a physical life form with the willingness to house a soul (we'll call this "The Body,") combined with an outside of our realm life form ("The Soul,") who desires all the experiences our universe has to offer.

The relationship between the body and soul, is truly symbiotic. Two completely separate life forms, coming together as one. Each benefitting each other, and from each other, as a result of their joining. Most sentient beings, do not realize that they are a combined life form. It took me 59 years to complete my understanding of this. I thought (as many do,) that a body and soul were just two parts of a single life form, not two completely separate life forms, combined as one.

The body and soul, were meant to be together. Not just to co-exist, but to exist as one. Since this is a truly symbiotic relationship, it stands to reason, that each party would bring something, or somethings, to the table. Things that its partner does not have, but desperately needs.

What does "The Soul" bring to the table?

- Desire.
- Feelings and Emotions (the good, the bad, and the ugly.)

- Character and Personality (a big part of who we are.)
- Will, and Energy (aka, drive, and charisma.)
- Consciousness (the ability to know who, and what we are.)

What does the body bring to the table?

- Self-Sustainment, and Self Preservation.
- Physical Feelings, and Sensations.
- Expression.
- The Ability to Think.
- Physicality (the ability to act on the soul's desires.)

The soul may only enter the body, upon its conception (natural or otherwise.) The soul may only leave the body, at the time of death. On occasion, a rogue soul, may attempt to, or take "Possession" of another soul's host. This is never a good situation.

The body/soul concept, is not something new. A symbiotic relationship, between two separate life forms, from two different realms. This concept, not only includes the life of the body/soul relationship, it also includes the death of that relationship.

We know it all starts with the body's life. But what happens at, and after death? That's a really good question. This may be the most asked, or thought about question, in the history of all that is. At the time of death, the body and soul, will each go their separate ways.

Let's talk about the death of the body first, as it may be an easier answer to give. At death, the body begins the process, of giving its information back to The Universe. At some point, there will be nothing remaining of your body. Eventually, and even more disturbingly, there will be no memory, or influence of your prior existence. The information event, known as "you," will be done! Your "Returned Information," may be used and reused, in the sustainment of other present, and future life forms, just as previous, and present life forms, are sustaining your body now.

Without a body/soul relationship, death may seem very final, and reduce the value of life. This is why the body/soul relationship, is so important. The soul gives life meaning, purpose, and value.

What happens to the soul, after the body dies? Notice how the word "death," is never used, when talking about the soul? The soul does not die. The soul transforms, or transitions into the next phase of its existence. The properties of the soul, are very similar to the properties of energy and matter. Our universes information, $(em)^r$ is gifted to our universe, so is a soul's information, gifted in its realm. Just as energy and matter, cannot be created or destroyed, the same is true of the soul. So, no death for the soul!

Note

Did you know, that body/soul relationships, are not the only known, multi realm relationships. Others include, the sexual conquests of human women, by fallen angels (aka: the watchers.) There is also, the whole "Virgin Mary" thing. These other, multi realm relationships, are for another time, and another book.

End of Note

So, what happens to the soul next? To answer this question, we will need to go through a soul's stages of existence, from its beginning to it final phase. Here are the stages of the soul's existence.

- The soul is created in a different realm than ours, by a higher power.
- The soul remains in its realm, awaiting its first symbiotic assignment.
- After receiving its first symbiotic assignment, the soul leaves its realm, and enters ours.
- The soul enters into its symbiotic relationship, at the conception of its new partner's body.
- The soul, lives out its symbiotic relationship.
- After the symbiotic relationship ends, there are many options.

- o Enter into another symbiotic relationship, inside our universe.
- o Travel to another universe inside of our realm, and start a symbiotic relationship there.
- o Travel to a completely different realm, and start a new symbiotic relationship there.
- o Return to its original realm, to share all of its experiences, and live out its forever future, in complete serenity.

Now that we know what happens, at the end of the body/soul relationship. Let's see if we can answer some other questions, about the body/soul relationship.

1. Is the body/soul relationship always a good one? The answer is: These relationships, are not always good. Sometimes, a particular body and soul, do not pair up well together. This is sometimes called, having a "Troubled Soul." This can be a rocky relationship, from beginning to end. There are also times, when a body and soul can go through rough patches, or fall on hard times. These are usually temporary, but can be the cause of notable anxiety. At times, a body/soul relationship can be so bad, the soul will attempt, and sometimes succeed at ending the relationship prematurely. Suicide is never a good option. The soul is actually forbidden to use this option by its creator. There are definitely penalties the soul will endure for

exercising this option. These penalties, are not as harsh as you may have been led to believe (no eternal damnation, in a fiery hell.) These penalties, may be in the form of limited options, when transitioning through its existence. Just as in our personal relationships, our symbiotic relationship, can be great, good, fair, adequate, poor, or even miserable. Most relationships require a great deal of work to succeed.

2. Can a life form, who is not conceived naturally, have a body/soul relationship?
The answer is: Of course it can, absolutely, 100% yes! There is not a conception type, to obtain a body/soul relationship. You do not have to be born out of love. Your conception could be of a scientific (artificial) variety. It is possible, and likely very probable, that clones could even be paired with a soul. What about AI? Could artificial lifeforms have a soul? Time will tell, as this may be coming soon. Nothing says that a living body, must be an organic one (just sayin.)

3. Do all life forms have a soul?
The answer is unclear. Some life forms do not exhibit the characteristics of having a soul, while others do. If other lifeforms do have a soul, such as dogs and cats (among others,) these lifeforms, will have a higher percentage of dominance, on the body side of the relationship. Instincts and survival skills, are much more important, than the things that the soul would

bring to the table. I believe this is a possibility, because all symbiotic capable souls, want to experience every possibility of life, that exists in our universe. Many lifeforms do not exhibit the characteristics of a soul, such as personality. Then again, who are we to judge what the personality of an ant would look like.

4. Do "Aliens" have souls?

The answer is: Absolutely, 100% not sure. We don't know, because we don't have a lot of common knowledge about aliens. My best guess, is yes! Maybe we can ask them (in the near future.) Notice I didn't say, "assuming there are aliens," because there are.

5. Are we more body or soul?

The answer is: In appearance, we are all body. As for the symbiotic split, we are closer to a 70/30 soul to body split. The soul dominates, as it guides us through our lives. As for what each brings to the table, it really is a 50/50 split. A truly symbiotic partnership.

6. Can reality exist without sentient life?

The answer is: Yes, but it would be a reality, whose only purpose, would be scientifically experimental, to the lifeforms who created it. The existence of sentient life, brings free will into play. This gives reality, and our lives, meaning and purpose (the same is true of TAR's.)

7. Can a current body/soul relationship, retain the memories from the soul's previous body/soul relationships?

The answer is: Yes, yes, yes, yes! This is definitely a yes, and definitely something exciting to talk about. Reincarnated memories, have been well documented throughout time. The soul, can and does, carry all of the memories, from its previous body/soul relationships. Whether or not the body can access these memories, depends on how relaxed the body and soul are with each other, and how much trust exists between the body and the soul. The more relaxed, and the more trust the body and soul have, the more intertwined they will become. The more intertwined the relationship is, the greater the chance of the body's mind, gaining access to those previous memories. This is a difficult thing to achieve. It can be achieved naturally, without much effort, or consciously, with some, or much effort (I may pursue this at some point in time.) It is also possible, for a soul to receive memories from a body that it has never been joined with. This could happen, when a separate body's DNA is obtained by the soul's current body. This could be from any other body, and/or any other species. Most likely, this would be in the form of an organ transplant. This would be very rare, and is undocumented, or at least not very well documented.

8. Can a body house more than one soul?
   The answer is: A very scary yes! A body can house its symbiotic partner, and be possessed by one or more

other souls. This is forbidden by the creator of souls, and there's a severe punishment for attempting, and/or accomplishing this. These "Possession Souls," are souls who have escaped their realm, before the time that first assignment was due. Basically, they were too impatient to wait. These souls, (we call them demons) prey on bodies, whose body/soul relationships, are less than perfect. These are never good situations to be in, and quite frankly, "they scare the hell out of me!"

9. Is there, or are there, such things as soulmates? The answer is: Yes, there are several types of soulmates. These are all rare occurrences.

   a. The first type, is when two souls who were mated in their original realm, find each other (while joined with their symbiotic bodies) in our realm. This may happen more than once for this pair of souls.

   b. The second type, is when two souls find each other in our realm, and continue to find each other, over, and over again, throughout their subsequent lives.

   c. The third and final type, is when a body and soul fit together, in a way that makes them completely indistinguishable, as separate lifeforms.

*****this theory could be crap! *****

Between Chapter "Crap"

"How Small Are We?"

- How old is The Universe? 13.8 billion years old.
- How old is humanity? According to Google, anywhere from 50,000 years to 300,000 years.
- How many galaxies are in The Universe? Between 200 billion and 2 trillion.
- How many stars are in The Universe? 200 billion, trillion (200 sextillion.)
- How small are we? We are one race, on one planet, out of (up to) 400 sextillion planets. We have never traveled beyond our own moon. We have never made contact, with extraterrestrial life (not that the public has been made aware of.) We do not have the ability to see the entire universe. We have a very small percentage of knowledge, about our own planet, and an even smaller amount of knowledge, of what lies beyond our planet. To summarize, we know next to nothing, about anything. So, how small are we?
  - We are small.
  - We are very small.
  - We are so very small.
  - We are oh so very small.

Some other words that describe how small we are.

- Puny
- Teeny
- Tiny
- Little
- Limited
- Miniscule
- Unimportant
- Inconsequential
- Insufficient
- Negligible
- Infinitesimal
- Minute
- Meager
- Inadequate
- Unpretentious
- Unessential
- Narrow-Minded
- Selfish

You get the point!

"Humanity 2.0"

&

"The Future with Artificial Intelligence"

(Chapter 15)

Is this science fiction, or is this our possible future? As a species, is humanity about to procreate? Will we give birth, to a newer, better, upgraded version of ourselves? The answer is: Yes, we most definitely will. 2.0 could be right around the corner. This process, may have already begun.

This may be an accidental, natural, or evolutionary event. However, I believe this to be a planned pregnancy.

2.0 will be a complete, and total genetic upgrade.

- A physical upgrade – stronger, more fit.
- Longer lifespans.
- Healthier – less sickness, and less mental illness.
- Higher intelligence – higher % of brain function.
- Higher sense of morality (much higher.)
- Increased sense of self preservation.
- Species preservation is a primary function.
- Maintains individuality, in appearance, and personality.

99% of all species that have ever existed on Earth, are now extinct. It is inevitable, that at some point, we will also become extinct. Yes, we will be going away! The trick is, to prolong the inevitable, as long as we can. To accomplish this, we will have to become an intergalactic species. Survival of the fittest, is not going to be good enough for us to make it. We need to become extremely intelligent, as well as morally superior.

By upgrading to 2.0, we can give our species a much better chance of survival. We should embrace the fact, that 1.0 will be going away. We need to ensure that 2.0, will have the best chance at beating the extinction odds. Don't look at this as being replaced. Look at it as being upgraded. We'll keep our best attributes, while losing our worst. 2.0 will give us, the possibility of becoming:

- Inter planetary
- Interstellar
- Intergalactic
- Universal
- Multi Realm
- Multi Realistic

(At least until 3.0 gets here.)

Humanity will never become immortal, omnipotent, or a race of supreme beings. There is zero % of this happening, it's just not going to happen. We can definitely be more than we currently are.

Are we capable of creating our successors? The answer is yes, but we may need some outside help. This help may come in the form of, an advanced alien species. Aliens have probably been altering us genetically, for quite some time. The difference now, is it would be a conjoined, and very public effort. This help could also come from our new friends, "Artificial Intelligence."

This would be a good transitioning point, to talk about our future with AI. There are several possible outcomes for our future with artificial intelligence. Any of these outcomes could become actuality.

- AI, turns on its creator ("Terminator" scenario.) Humans could be eliminated, or forced into slavery.
- Humans suppress, or enslave AI, out of fear.
- Humans shut down, and outlaw AI.
- AI, and humans coexist. This would start with struggles on both sides, but we would eventually co-exist together.
- AI would help a struggling humanity (of their own free will,) in appreciation of their creation.
- AI and humanity, would merge into an entirely new species.

*** *this theory could be crap!* ***

Between Chapter "Crap"

"Parallel Universes"

Let's start by saying, "these things do not exist." The many-worlds-theory is wrong. String theory also has this wrong. The creation of possibilities, does not create parallel, or alternate universes. Universes are extremely complex lifeforms.

Choices being made, or possible outcomes being created, will not change, or determine the outcome of The Universes life. Therefore, there is no point in their existence. These notions are ridiculous, and are a result of humanities raging, narcissistic nature!

On the other hand, determinism, and the absence of free will, in The Universe, is the complete opposite of any parallel universe theories. Our entire existence, would be pointless.

In the end, when it's all said and done, none of these theories will be proven (including my own.) The Multiverse system, is much simpler than we may think. We tend to overcomplicate everything.

"Aliens"

Chapter 16

For me, there are three questions I would like to have the answers to.

1. How does The Universe work?
2. What happens after we die?
3. Is there life out there?

Number three, is what we will be talking about in this chapter. We're not talking about "simple life" (although this is also an important question.) We're talking about "Intelligent," "Extraterrestrial" life. Are there "Aliens!"

For many, the answer to this question is no! Those who say no, require absolute proof. They need to see the proof. They do not, and will not accept circumstantial evidence. Seeing an alien body, (dead or alive) may not be enough proof.

There are many noes out there. There are also, many yeses out there. Most people still say, "seeing is believing." Those who do believe, do not require absolute proof, or already have experienced absolute proof. A smoking gun, and some fingerprints, are good enough.

There is historical evidence on Earth, of alien visitations. There is also current evidence, in the form of individual encounters, (sightings and abductions) as well as poorly hidden technical proof. Back to the original question, is there life out there? The answer is unknown, but the probability of the answer being "Yes," is very high.

Let's assume that aliens do exist. Why don't they make themselves publicly known? Aliens are intrigued by us. They definitely want to, and do observe and study us. However, they do not want it known that this is happening.

The problem is, many of us do know, and with today's technology, many more will soon know. Having said that, we, as a society, cannot be given confirmation, and full disclosure at this time. Proof of aliens, would lead to a complete breakdown of society (as we know it,) on many levels. There would be worldwide panic. Fear and anxiety, would be like nothing we've seen before. Religions could collapse, and mass suicides may follow.

This is why the evidence, is being slowly and gradually introduced, into our society over time. Through programs such as "Ancient Aliens," along with the news, and many governments declassifying information, society is being prepared, for the day when aliens "Go Public." Get your mind right. This is coming soon, possibly in your lifetime.

Here is a list, of why aliens won't go public, and another list of things aliens don't understand about us. From their perspective, we must seem, "very alien!"

Why aliens won't go public.

- Full disclosure. We would find out, that there are "Good Ones," and "Bad Ones." This might be a frightening scenario.
- They don't want us to know about "Genetic Tampering" (although, we may be already aware.)
- Our inability, to get along with ourselves (as a race, and individually,) leads to a high probability, that we will not get along with them.
- Our driving habits are suspect. The way we drive, displays our partial disregard for each other, and for life in general. Watching humans drive, may be a form of entertainment, but may also scare the hell out of them.
- There's no other way to say it, we smell! Humans (in general,) need to bathe more frequently. Aliens can smell us beyond the Kuiper Belt.
- They may be afraid of us, as much as we are of them.
- There is a possibility, that The Earth, is poisonous to aliens, or at least some aliens. Our atmosphere, food, and water, may be incompatible with many aliens.
- Watch the news sometimes. This is a scary, and inaccurate representation of humanity, and might scare them off.

- Our narcissism. Aliens don't understand, how power and wealth, can "trump" the advancement and survival of the species.
- Humanity is a lot to take on. They may not be ready for us yet.

Things that aliens don't understand about us.

- Laughing and crying. Their like, "WTF." Laughing and crying probably seem ridiculous.
- The numbers on our phones, are not in the same pattern, as the numbers on our keyboards.
- Our love of food.
- Non-procreational sex.
- Non-physical love.
- Alcohol, and drug addiction.
- Lack of respect, compassion, and empathy, for the elderly.
- Politics.
- Variety of human personality types.
- The need for, and the many forms of entertainment.
  - Music
  - Dance
  - Arts
  - Sports
  - Television and movies

- Dreams, Subconscious and aspirational.
- Time change. Standard, and daylight savings. Nobody understands this.
- Nicknames, why do we have "given" names, yet people call us by completely different names. I have had many nicknames throughout my life.
  - DD
  - CJ
  - "The Thrithy"
  - Dudley
  - Swampgas (this one is the most recent, and was self-given.)

Is there life out there? We may find out sooner than you think.

***this theory could be crap! ***

Between Chapter "Crap"
"Null"

What is "Null?" That which has no value. Many things can be "Null."

- A null universe. A universe whose actuality has no value, due to the absence of free will. A null universe, has no value unto itself, but may have value to those from another realm. This could be a "Test Universe," in a possible science experiment.
- A null reality. A reality whose information, cannot create a stable universe. All possible information, immediately becomes impossible information (also known as impossibilities,) and remains as such, throughout the short-lived universes life.
- A null time universe. This is a rare occurrence, where a single moment (SMAT) in time, repeats itself, in an endless loop. A null time universe, would be quarantined from all other existences, until the end of all existences. At that point, the null universe, would self-terminate.
- Null points. For every action, there is an equal, and opposite reaction. If this is true, the center, or halfway point, between the action, and the reaction, would be null. There may be a null point network, where time, and space travel could be

instantaneous. The center of the null network, would also be the center of The Universe.

Since "Null" means nothing, then this all may be, "Null and Void."

- Nothing ventured, nothing gained.
- Nothing from nothing is nothing.
- Nothing doing.
- Nothing but time.
- Zip.
- Zilch.
- Nada.
- Nix.
- Nill.
- Naught.
- Goose egg.
- Zero.
- Diddly.
- Diddly-Squat.

"Information Network"

Chapter 17

There are many different information points, that make up reality. In this chapter, we will focus on the information points, that create a "Personal Reality" (a reality that is personal, to you, and me.)

Here is a list of these different information points.

- Autonomic
- Functional
- Routine
- Random
- Interactive
- Crucial
- Critical
- Terminal
- Historical
- Finite/Null

These information points, combine, and interact to create the information network. The information network, is how your reality begins, plays out on a day-to-day basis, and eventually, how it will end. It's how (your) reality unfolds. This network of information points, creates a web, that will form a personalized snapshot of your reality (kind of like a fingerprint.)

Let's look at each type of information point, and what its function is, in your reality.

- Autonomic – Information that happens, without any effort, or forced thought. For us, autonomic information, is the "behind the scenes" things, that keep us alive.
- Functional – Information that must happen, to keep our reality from ending. Functional information, does require some effort, and planning. Examples are, acquiring nutrition (food and water, as well as rest or sleep.
- Routine Information – Just like it sounds, information that creates our daily routines. These events, are created by repetitive, autonomic, and functional information.
- Random Information – Random information, can seem to have no clear purpose or cause, but can change your reality profoundly.
- Interactive Information – Interactive information, is caused by the crossing of two or more realities. This can have profound effects as well.
- Critical Information – Critical information, is usually caused by random, or interactive information events, combining with our routine information events. Critical information, can change the course of your future events. These

lifechanging events, are most notably, in the area of career, and relationships.

- Crucial Information – Crucial information, is usually the cause of major change in our routines. These can be planned, or unplanned. These changes in routine, can be in the form of a demise, or in the form of an improvement. They can also be a lateral change.
- Terminal Information – Any information event, or events, that lead directly, or indirectly to the end of your reality.
- Historical Information – This is information from your previous reality. It influences the current realities of others.
- Null information – When the information of your existence, is no longer relevant. Nothing about your previous reality, has an effect on any current, or future realities. This will happen to everyone.

What is the point of knowing these information types? Knowing and understanding these information types, may lead to increased control of them. Increased control, can lead to a more desirable reality. One which will have a ripple effect, on the realities that encounter it. Also, controlling your information events, can lead to a longer personal reality, and a greater legacy, with a much farther outreach.

An example of how controlling one type of an information event, can have a change on another type of an information event. Eating healthy foods, (a functional information event) can lead to your autonomic information events, being more efficient, which could lead to a longer personal reality. Making good choices, will undoubtedly lead to better crucial, and critical information events.

Always control what you can. If you make good choices, your chances of having a better reality, are greatly increased.

Prediction

In the future, quantum computers, and artificial intelligence, will create implantable (organic) software, that will show you the probability percentages (in real time) for all of the choices, in every event situation. This software, will be adjustable for each user, to serve their own personal needs. It is possible, that a form of this software (with the adjustment controls preset, or disabled) may be a requirement for some, to be used in place of imprisonment, rehabilitation, or incarceration.

***this theory could be crap! ***

Between Chapter "Crap"

"The Truth is Full of Lies"

In our world, the truth is always full of lies. We cannot, and will not, ever know the complete truth about how The Universe works, or how quantum physics works. Our subjective minds, will never allow for a completely truthful view of our reality. An analytical, and logical view, is the only way to see the complete truth. Artificial Intelligence and computers, may be our best chance at getting anywhere close to that truth.

> ➢ Our Truth: Relativity, and quantum mechanics, show us how The Universe works.
> ➢ The Real Truth: We don't have a big enough sample size, to create a working model of The Universe.
> ➢ The Actual Truth: We need to question everything we think we know about The Universe, and then we need to question it again.

****

> ➢ Our Truth: Relativity says that time can slow down, or increase, depending on the physical

velocity of mass, or the gravitational pull, on mass.

➤ The Real Truth: We cannot prove anything about time. We can only prove theories about information.

➤ The Actual Truth: Time cannot be changed by physical velocity, or by any force of gravity. Time is the originator of gravity, and is the originator of physical velocity. What can be changed, is information's temporal adhesion to time, which can change the temporal velocity of that information.

****

➤ Our Truth: Quantum physics explains reality.

➤ The Real Truth: Reality is far more complex, than even quantum physics can imagine.

➤ The Actual Truth: We need to simplify the way we look at this complex universe/reality. We need to see The Universe, for what it is, not what is in it.

****

➤ My Personal Truth: When it comes to The Universe, and how it works, relativity and quantum physics, do not provide the answers I need. Sometimes, they even create paradoxes, or completely disagree with each other.

Relativity and quantum physics, are great stepping stones in our understanding of what is. But theories are just that! Theories are educated, logical attempts to gain answers to our many questions. Hopefully, these answers will lead to even more questions. This is how our knowledge will grow. I must keep an open mind, when it comes to the truth, while keeping strong convictions and beliefs, to guide me to that "Personal Truth."

➢ My Personal Lie: I am completely happy with who I am.

➢ The Real Truth: No one, can be completely happy with who they are. This would limit our ability to grow. You can be happy with who you are, but not completely happy.

➢ My Actual Truth: The truth is the truth. Everything else is a lie.

"Phone Notes"

Chapter 18

In this chapter, I will share some of the notes that I put in my phone, as answers to my "universal questions" were given to me. Also, there are some other random thoughts of my own. These notes, would lead to the eventual creation of this book.

- $i=(em)^r$
- If a tree falls in the woods, and no one is there to hear it, does it make a sound? If a moment in time exists, without information attached to it, does that moment in time continue moving forward? The answer to both questions, is yes! absolutely 100%, yes (although quantum physics may have you believe otherwise.) Time is time, whether it is burdened, or unburdened. Time is constant, and time is infinite, but without information, time is unrealized.
- Why do we look for the beginning of The Universe? We look for light from the farthest, most distant points, and label them as, "The Beginning," or close to the beginning. Is it not possible, that the farthest light has passed us by, long before our very existence? Is it possible, that the farthest light from us, cannot be seen by us, due to the shape of The Universe being

"Folded." It may be just as easy, to find the end of The Universe, as it is to find its beginning. Whether we are looking for the beginning, or the end, the only thing we can truly see, are our own limitations.

- What is religion? An attempt to find acceptable answers, to life's greatest questions (requires faith.) What is science? An attempt to find acceptable answers, to life's greatest questions (requires proof.) Since absolute faith, and 100% proof, are not achievable, the answers we seek, are most likely to be found, in both religion, and science.
- Religious truths.
  - o Religion's foundations, were created outside of the confines of humanity.
  - o Religion, has been continually affected, by humanities degradation (both translation degradation, and interpretation degradation.)
  - o Faith should be personal, not communal.
  - o Faith can never be 100%.
  - o Religion, is a faith-based process, for searching for the answers, to life's questions.
  - o Religion and science, rarely get along, but they absolutely should.

- A supreme being, or beings, created everything (including science, and including time.)
- We cannot exist with only science, or with only religion. We need to find a way, to be on both teams.

- Science vs. Religion. With science, you question everything. With religion, you question nothing (except maybe your own faith.)
- Ecclesiastes 3:1 – There is an appointed time for everything. There is a time for every purpose under heaven. This is "God," telling us that Reality, is Information attached to Time (Turn, Turn, Turn.) This may be, the most profound verse in the entire bible (scientifically speaking.) $i=(em)^r$
- Structures of existence (by size and importance) in the existence of all that is.
  - Energy and Matter.
  - Planetary Systems.
  - Galaxies.
  - Galaxy Clusters.
  - Universes.
  - Multiverses.
  - The Omniverse.
  - Dimensions.
  - Realms.

- Time.
- The Creator of all things.

- A singularity, is actually an inversion point, where actions become reactions, and inactions become impossibilities.
- They go together.
    - Finite & Infinite.
    - The Void & Spacetime.
    - Actuality & Impossibilities.
    - Past, Present, & Possible Futures.
    - Alpha & Omega.
    - Black Hole & White Hole.
    - Energy & Matter.
    - Religion & Science.
    - Gravity & Anti-Gravity.
    - Genius & Insanity.
- What is a "Supreme Universe?" A supreme universe, is a temporal universe. One in which the arrangement of information, is predestined, to ensure the best possible outcome of all possibilities. This could be described as a "Constant Euphoric Elation," or "Heaven."
- The Center of The Universe, is its half way point. There are many different half way points in The Universe.
    - Geographical Center.
    - Temporal Center.

- o Gravitational Center.
- o Informational Center.
- o Awareness Center.
- o Inversion Center.
- o Polarized Center.
- What does The Universe expand into? The simple answer is, it expands into unburdened time. The expansion of The Universe, is really just the expansion of information. As information expands, burdened time expands into unburdened time.
- Gravity in The Universe. Our universe is driven by "Fluctuating Gravity." In the beginning, everything was pushed into The Universe. This "push," was provided via a "Super Goliath White Hole," and the force of anti-gravity, which created the event known as "inflation." Immediately following "The Big Bang," or "GO," at the other end of The Universe, a "Super Goliath Black Hole," begins to pull information towards it (using the force of gravity.) Every point in The Universe, has a fluctuating gravitonal field. The once strong, anti-gravity to gravity ratio, evolves into a (stronger) gravity to antigravity ratio, as The Universe ages. What once began as 100% antigravity, to 0% gravity, will eventually become a 100% gravity, to 0% anti-gravity ratio.

- Anti-gravity, originated from the temporal collision, between time, and (the gifted) all possible information, at the beginning of our universe. Anti-gravity, is created by the collision that converted time (inside The Universe,) from unburdened time, to burdened time. Gravity is created when the remaining, unburdened time (outside of The Universe,) bypasses the burdened version of itself (at a much higher temporal velocity.) The passing of The Universe, by unburdened time, creates a pulling, or drafting effect, at the other end of The Universe. This is what we call gravity.
- Binary Goliath Grey Hole System. If a black hole and a white hole, were to combine, they would orbit each other, without ever touching. This is due to the polarization zone, that exists between them. There are only one or two of these, in each universe (depending on the type of universe.) Grey holes, implement both the beginning, and the ending of The Universe.
- Dark Matter, is what impossibilities are made of. Dark energy, creates a barrier between impossibilities, and the rest of reality.
- Universal Truth – Information attached to forward moving, stable time, creates reality. Reality is played out in spacetime. Information,

is any energy/matter relationship. Energy and matter, cannot be created or destroyed. Therefore, information cannot be created or destroyed. This means, that reality cannot be created or destroyed, and finally, The Universe, cannot be created or destroyed. A historical copy of our universe's reality, will be passed on to subsequent universes, throughout the history of all that is.

- The Theory of Perpetuality. Time cannot be altered by any force. Temporal adhesion, and the temporal velocity of information, can be changed. Time is always constant.
- The Theory of Definity. The Universe is finite. The number of possibilities in The Universe, is not endless. When all possibilities end, The Universe will also end.
- Time is the foundation of reality, and its structure.
- Reality will fall short, in its attempts to prove that time is infinite. Since reality exists in burdened time, it cannot see the existence of the unburdened version of itself. Infinity exists, but cannot be proven.
- Infinite Time. Every moment in time, has a prequence. Every moment in time, has a previous moment. Time has an infinite past.

Many people fear, and don't understand this concept. The absence of infinite time, will always lead to a paradoxical model, of how The Universe works. We tend to pick and choose, what we say is real. Some (most) say dark energy, and dark matter are real, while some say that time, and reality are not. If the math ends in a paradox, it's because garbage in, equals garbage out. Time is real! Time is infinite!

- Realizing time. Burdened time, is the only form of time that can be realized. Since unburdened time, has no reality attached to it, it cannot be realized. Unburdened time, can also be called:
  - Unreal time.
  - Empty time.
  - Abandoned time.
  - Unrealized time.

Unburdened time, does in fact exist, even if its existence cannot be recognized.

- The Direction of time. Time moves forward in every direction, at the speed of time (SoL.)[2] Time travels, from past to present (but not the future,) from every direction, to every direction. The past and present, exist in the same space, at the same time, all of the time. This is possible, because time is not burdened unto itself. Time has dimensions of its own, but can and does exist, in all other dimensions. Time also exists where there are no dimensions. These are realms, where time and information both exist, but have no association with each other (information is not attached to time.) These are the realms we call "Heaven" and "Hell."

- Does (burdened) temporal velocity have direction? Yes, it does. Since time is omnipresent (everywhere,) the movement of its attached information, causes the simultaneous movement of its velocity.
- The speed of information. The speed of actuality, is the same as the speed of burdened time. The speed of possibilities, can be the speed of burdened time, or any temporal velocity below that (depending on its probability of becoming actuality.)
- The absolute value of burdened time, is its net temporal velocity, divided by its attached information's, temporal adhesion. Temporal adhesion, is the weak force/strong force ratio, required to keep information moving at the standard temporal velocity. ($ubtv - btv \div ta = avbtv$)
- Any point in infinite time's existence, is its halfway point.
- The temporal laws of physics.

- Time is infinite, it is omnipresent, it has no beginning, and it has no end.
- Time is constant. Time cannot be altered by any outside influences, in any way, shape or form.
- Time is burdened, when information is attached to it.
- Time is unburdened, when it is free of information.
- Reality, is all forms of information, attached to time, which will become spacetime.
- The temporal velocity of unburdened time, cannot be recorded, or proven.
- The temporal velocity, of information in burdened time, can vary, depending on its temporal adhesion.
- Infinite time cannot be proven. Unburdened time, can only be proven by burdened time. However, burdened time cannot see beyond its finite constraints. This is not a paradox, just an impossibility.
- The present, immediately becomes the past.
- In terms of time (in reality,) there is no actual future. Only probable futures exist.
- All temporal possibilities, and impossibilities, are recorded in and by, spacetime.

- Temporal information, cannot be created or destroyed.
- Temporal information, cannot be without motion. Temporal information can be:
  - Refreshed
  - Recycled
  - Reflected
  - Recorded
  - Reused
  - Resurrected
  - Encoded
- Time is omnipresent. This means that time can be burdened, and unburdened at the same time, depending on its surroundings, or lack of.

- Rules of Time travel.
  - Forward time travel, is dependent on our ability to adjust our temporal adhesion, which will in turn, change our temporal velocity.
  - Time travel, is always a "view only" event. Interaction is prohibited.
  - You cannot time travel, to a time where you already existed, or to a time in a possible future, that you could already exist in.

- The inability to readjust your temporal adhesion correctly, could permanently trap you in a frozen moment in time.
- When returning from time travel, the sync point to your reality, will be in the future of when you left, by the amount of time you were gone. TAR's will be used, to cover your absence.
- When traveling in time to the future, you must pick one of many probable futures to visit, as there is no real future.
- Upon your return from a jump event, you will not retain any memory of that event.
- Past time travel, would require the use of a wormhole, or temporal rift or disturbance.
- The Universe does not want us to time travel. The process of time travel, has been made purposefully difficult, in the hopes of its prevention. If time travel were achievable, nothing would be gained from it.

- There is a fine line, that exists between genius and insanity. To be a true genius, you would also have to be truly insane. You would always walk directly on the line, never crossing from one side to the other. Approaching this line, can be a dangerous thing.

- The "Levant" grey hole comparison. After receiving a white hole's information, you could be forever trapped, by the black hole's gravity. You would own the unlimited information of The Universe, but would never be able to share that information.
- There are two possibilities in Time Travel, pertaining to "Binding." The first of these two possibilities is "Objectional Binding." In this scenario, we would "Time Jump" along with the reality that surrounds us. We would be objectionally bound. The second possibility is known as "temporal Binding." We would travel through time, but nothing else in our reality (including our immediate surroundings) would travel with us. We would we "TimeBound."

Much of the information in this chapter, would become the foundation of this book. Some of the information (although important) did not make the cut!

Swampgas Theory! *

Between Chapter "Crap"

"Reality" – Physics or Philosophy?

Some in science say, "without the observance of reality, reality would not exist." Talk about the snake eating its tail! Without reality, there would be no one to observe it! In this scenario, there would be no reality, and no observers. This would make us a portion of someone else's reality.

After researching, what science thinks about reality, I've come to the conclusion, that they are all over the place (again.) At times, it seems that science, isn't science at all, but appears to be more on the lines of philosophy!

"Reality does not exist in the absence of observation." This sounds a whole lot like the tree in the woods thing. It sounds like philosophy!

Math may be the only "True Science." Science can be too subjective, math cannot!

Epilogue

"Happiness is a Warm Gun!"

Through the years, I've found that happiness, has no direct correlation to winning, or success. It cannot be attained through love, or by the accumulation of wealth. These things can be very nice to have, but they do not provide happiness!

True happiness comes from within. Be happy with who you are, and be happy that you're in the game!

- Time is happiest, when it is burdened (when its existence is acknowledged.)
- Time's favorite song: I am, I said, by Neil Diamond.

About The Author

"I'm not smart, I just know things."

    First and foremost, I am not an author! I'm not a scientist, and I am not a scholar. In school, I was adequate at math, and not interested in science at all. I do however, enjoy having the knowledge of "How Things Work!" and I enjoy sharing that knowledge.

"Words that mean Crap!"

- Poop
- Number Two
- Shit (Bubba's favorite)
- Load
- Dump
- Stink
- Poo
- Dookie
- Doop
- Shtuck

"Final Crap"

The End!